Marcel KOUROUMA

Governance of common resources in the Ziama classified forest

Marcel KOUROUMA

Governance of common resources in the Ziama classified forest

Participatory approach: an alternative for better management of common resources in the Ziama forest reserve

ScienciaScripts

Imprint

Cover image: www.ingimage.com

This book is a translation from the original published under ISBN 978-620-6-72456-8.

Publisher:
Sciencia Scripts
is a trademark of
Dodo Books Indian Ocean Ltd. and OmniScriptum S.R.L publishing group

120 High Road, East Finchley, London, N2 9ED, United Kingdom
Str. Armeneasca 28/1, office 1, Chisinau MD-2012, Republic of Moldova, Europe
Printed at: see last page
ISBN: 978-620-8-32954-9

Contents

DEDICATION

I dedicate this book to my mother MAMY ROSALIE THEA who gave me an education of rigour and humility from my earliest childhood. Despite her meagre means, she also helped me to gain access to education, which is illustrated by the writing of this memoir. I would also like to pay tribute to my father ROGER KOLY KOUROUMA, who has always made me believe that I can do anything and has kept my hopes alive to this day.

I pray that Almighty God will grant them long life.

Resume

The classified forest of Ziama covers twenty-eight (28) villages divided between five sub-prefectures, of which Seredou is the largest, having most of its territory in this forest through the villages of Singbedou, Sanssanconie, N'Zebela and Oreme.

While all these villages face similar difficulties due to their geographical location in relation to the Ziama forest massif, those in the enclaves are in a more critical situation.

These villages are located inside the forest and therefore have no cultivated land outside their village. However, it should be noted that the main activity of these populations is agriculture. This agriculture is marked by the use of rudimentary tools that are less suited to satisfying the subsistence needs of these populations.

During the classification of the forest, village territories were defined according to the number of inhabitants in each enclave at the time. In 1942, this policy was probably a reliable solution for conserving the massif. But, years later, the population of these villages had increased very rapidly, and the space defined by the colonial administration was becoming insufficient. Hence the local people's desire to have the boundaries of their land changed so that they could spread out their farming activities.

This led to a conflict between the forestry administration and the village communities. To this day, this conflict makes it difficult to manage the forest in the enclaves.

The enclave method[2] was certainly a solution at the time. However, with the rapid growth in enclave populations, this policy needs to be restructured.

[2] First of all, enclaves are villages located in the middle of a classified forest and which have their land within the forest. For the colonist, a way had to be found immediately to enable the people of these villages to live on their land, but this time with a limit. This meant defining boundaries according to the number of people living there at the time.

INTRODUCTION

The sustainable management of forest resources is now a real issue. Since the 1992 Rio de Janeiro climate conference in Brazil, the forest issue has remained at the centre of international debate. Because of the particular attention paid to this resource, as a biotic reservoir and climate regulator, all the nations of the world are now concerned with its sustainable management through projects to maintain its balance. In Rio, a declaration on forests and measures for their conservation was adopted, based on the concept of sustainable development. Ten years later, in 2002, a climate summit was held in Johannesburg, at which, in line with the Millennium Development Goals, UNCED adopted a declaration on forests and set up the United Nations Forum on Forests (UNFF). Once again, the forests and the people whose livelihoods depend on them were the focus of attention. Since then, the Member States of the United Nations have implemented the Rio Agenda21 and international conventions enabling humanity to move towards sustainable forest management. Since then, a number of initiatives have been taken, including the financing of concrete actions to protect forest areas. These two major conferences subsequently put the forestry issue at the heart of environmental discussions.

At the Nagoya conference in 2010, the Convention on Biological Diversity (CBD) advanced the cause of protecting forests as a reservoir of biodiversity. To this end, a ten-year action plan was adopted, which calls for the rate of loss of natural habitats, including forests, to be reduced by at least half, if not to almost zero.

To give the forest even greater importance, the United Nations system declared 2011 the "International Year of Forests" at its General Assembly.

Since Rio in 1992, through Johannesburg in 2002 and Kyoto in 2012, numerous efforts have been made in the area of natural resource management in general and forestry in particular, resulting in considerable progress in the written word. Alain BILLARD (2013) points out that governance in the forestry sector has particularly evolved at supranational and international levels since Rio. The various studies carried out in the forestry field emphasise the empowerment of communities at grassroots level, as emphasised by Nathalie RIZZOTTI (2004),[3] . The international context of the 1990s led to the strengthening of partnerships between conservation agencies and local populations, thereby promoting a new

[3] Nathalie Rizzotti, coordinator of the UNESCO Chair in Lausanne, in her book "Technology for sustainable development".

The United Nations Conference on Environment and Development, held in Rio de Janeiro in 1992, was the starting point for the focus on forests as renewable natural resources and carbon sinks (such as the Amazon rainforest).

approach to resource conservation known as 'community conservation projects'. An approach that envisages a variable management method and also the establishment of an institutional status for areas where management is carried out by the communities themselves.

Joint resource management is based on the observation that communities can be better placed than the market or the government to manage the resources of their localities, their everyday environment. This local control leads to a more respectful or sustainable relationship with the environment.

Natural resources are the source of the wealth that enables us to improve the living conditions of the world's populations. Despite their abundance, they are subject to very strong human pressure, which is increasing from year to year because of population growth, to which are added the effects of inappropriate management and exploitation methods. The fact is that the development model advocated by nations is not respectful of the environment, and this has led to an imbalance in ecosystem-society relations as a result of the frantic search for wealth (minerals, firewood, charcoal). This quest is compromising the productive and regeneration capacity of the forests.

Faced with this situation, the international community, sensing the threat of a global crisis linked to the depletion of natural resources, began to reflect following the holding of several international events, such as the United Nations Conference on the Human Environment (UNCHS/Stockholm-1972) and above all the Conference on Environment and Development (UNCTAD/Rio de Janeiro -1992), with a view to establishing global governance of resources, taking into account all the players involved. This conference gave rise to the conventions on climate change, the conservation of biological diversity and the fight against desertification. All these post-Rio conventions aim to adopt a global approach to the conservation of biological diversity, the sustainable use of natural resources and the fair and equitable sharing of the benefits arising from their exploitation. Thus was born in 1971 the notion of sustainable development, which today is understood as a process of development that respects the environment and ensures a better balance in space-social relations through an appropriate symbiosis of the biophysical, human, socio-economic and institutional dimensions .[4]

In the wake of Rio, a number of countries have signed up to the principle of sustainable management of their natural resources. Like many countries around the world, the Republic of Guinea was also actively involved in drawing up the Rio Convention on Biological Diversity, which it signed in June 1992 and

---

[4]*(synapse.uqac.ca/wp../**Forets_and_humans_Complete_Study_Chap_05**.pdf)*, site consulted on 04 April 2015

ratified on 07 May 1993, becoming the second African country and the sixteenth of all the contracting parties [...]. In order to comply with Article 6 [5]of the Convention, Guinea has undertaken a national assessment of its biological diversity (MTPE, 1996).
Following ratification, Guinea made a firm commitment to the sustainable management of its natural resources.
However, given that 72% of the population lives in rural areas and relies exclusively on agriculture, forest resource management remains a major concern. It has an estimated population of *11,672,876* inhabitants Daniel LAMAH (2013), with an annual growth rate of 3.1% [.] this estimate corresponds to a demographic density of 47.48 inhabitants/km2, which remains low overall. 45% of this population is under 15 years of age, and those living in rural areas in the same age bracket account for 48% of the total, compared with 38% in urban areas. The working population represents 50.5% of the total. Guinea is one of the poorest countries in the world. According to Daniel Lamah (2013), the incidence of poverty is currently 55.2%. At the limit of their survival, these populations are therefore engaged in the destructive exploitation of natural resources through the extension of crops on marginal land by clearing land in search of new land. These means of subsistence sometimes have unfortunate consequences for natural resources.
These figures make it difficult, if not impossible, to manage the country's classified forests, most of which are located in rural areas.
To achieve this, it is also necessary to develop community or participatory management systems, although the Rio Summit recommended that signatory countries transfer the power to manage forest resources to local communities. This management experiment has been successful in countries such as Zimbabwe, Mali and India (Patrick TRIPLET, 2009). This would appear to be a favourable context for giving local people greater responsibility for managing natural resources. We believe that this summit marked a new turning point in forest resource management, enabling rural populations to become directly involved in forest resource management while benefiting from the financial spin-offs.
The Ziama forest, our study area, is the country's largest forest reserve, with a rich and diverse ecosystem and, what's more, a level of degradation of 27%, which appears to be the least extensive of all the country's forest reserves, and is even considered "virgin" by some authors.
Like other protected areas in the world, the Ziama forest has two particular

[5] This article stipulates that "each Contracting Party shall develop national strategies, plans and programmes for the conservation and sustainable use of biological diversity".

problems that are described by protected area representatives as damaging [...] the practice of itinerant slash-and-burn agriculture, which is seen as destroying the vegetation, and hunting for the meat trade (poaching). In contrast, the main preoccupation of the villages bordering this reserve is confronting the elephants that ravage their food fields.

This memoir comprises three chapters;

In the first chapter, we present the problem, the methodological approaches, the conceptual and theoretical tools and the analysis model. The focus is on the issue of natural resources. We focused on three key concepts for our research.

i- the concept of resources; ii- the concept of governance and; iii- the concept of stakeholders. To illustrate these concepts, we have drawn on the writings of numerous researchers around the world.

The second chapter deals with the Ziama forest massif as an ecosystem with diverse and complex issues.

In the final chapter, we present participatory management as an alternative for rational and sustainable protection of the Ziama forest ecosystems.

We will end this report with a conclusion and a perspective for the thesis.

CHAPTER I

PROBLEM AND METHODOLOGICAL APPROACH

1.1- Problem

The Gurnee, a West African country, belongs to the tropical zone due to its position in longitude and latitude. It is characterised by the existence of forests in its four natural regions, with climatic and plant diversity. According to the French Ministry of Ecology and Sustainable Development (MEDD5, 2009),[7] , Guinean forests cover some 13 million hectares (53% of the national territory). It is divided between 250,000 ha of mangroves; 700,000 ha of dense humid forest in the south-east; 1,600,000 ha of dense dry forest and open forests in the north and 10,636,000 ha of wooded savannah. There are 162 classified forests covering a total area of 1,182,133 ha (5% of the country). Almost all are located on mountain ranges, and these classified forests help to protect water catchment areas and their springs.

In Gurnee, ownership of the forests belongs to the State, which delegates a right of use to the users, in accordance with the terms and conditions defined by the relevant legislation in force (Pascal REY, 2007). However, the texts recognise the customary rights of local populations if the use of forest resources by these populations is limited to satisfying their basic needs. Local people's day-to-day relations with forests are therefore governed by social norms that come under customary law. This customary law confers jurisdiction over forestry issues on traditional chiefs, who delegate responsibility for managing each component of the natural resources to a patriarch, or to the heads of clans, lineages and families (Pascal REY, 2007).

However, the forests are under the supervision of the Ministry of Water and Forests or the Environment, which decides on the allocation of forests to local populations in collaboration with other ministries concerned with decentralisation or the management of other resources dependent on the health of the forests. The texts also provide for the involvement of state actors at various levels (governors, prefects, sub-prefects, mayors, etc.). In addition to ministries, local administrative authorities and local populations, non-governmental organisations and associations also play a very important role in implementing the decentralisation of forest management. However, this decentralisation of forest resources is not being carried out in an inclusive and participatory way. This shows the limits of decentralisation in forest resource management. In our view, this decentralisation needs to be improved. The

[7] Ministry of the Environment and Sustainable Development.

marginalisation and exclusion of local people, and the failure to take account of their customs, are not conducive to the concerted involvement of local people and their total support for such a policy. This exclusion of local people from decision-making appears to be a weakness of the system. This situation prevents the objectives being pursued from being achieved.
As we can see from this FAO report published in 1994, which estimated annual deforestation at 1.3%, or 87,000 hectares. In 2003, Marcel SOW estimated the annual rate of deforestation of Guinean plant cover at between 60 and 70%. These figures seriously compromise the future of the Guinean forestry sector. The Guinean environment is not only a national heritage, but also an integral part of the universal heritage (the Monts Nimba Biosphere Reserve), which provides mankind with resources. Its conservation is essential to maintaining climatic equilibrium. As a result, the management of this heritage must take account of all stakeholders at all levels.
The country's dense rainforest ecosystems are located in the south-east of the country, in the region known as Guinee Forestiere. They are of the Libero-Ivorian type. Once covering 14 million hectares, only 700,000 ha remain, divided between the Prefectures of Gueckedou, Lola, Macenta, 'Nzerekore and Yomou (Amadou BAH et al, 1996). This region contains the country's largest forests, due to the presence of a dense secondary forest. However, all these forests are undergoing extensive degradation. These forests include :
- The Diecke forest covers 59143ha and is located at an altitude of 400-595m. It is a wet lowland forest with 30% degradation;
- Dere 8920ha floodplain forest at the foothills of Mt Nimba with a level of degradation of 90%; Mt Bero 26850 ha located at an altitude of 600-1200m, a semi-deciduous forest and savannah with a level of degradation of 75%;
- Pic de Fon 25600ha located at an altitude of 600-1656 m; it is a humid mountain forest, with a level of degradation of 35% caused in particular by the exploitation of wood, the forest of Ziama with a level of degradation of 27%.
The majority of protected areas are in wetlands, although conflicts over the use of resources in these areas, due to growing demographic pressure, have limited their extent. Vast areas have the status of classified forests, but in reality they are being left to their own devices and are suffering severe degradation to varying degrees. We believe that these areas should benefit from a higher level of protection to prevent their disappearance.
The Ziama forest which is the subject of our study is one of the classified forests of the region since 1942). It was classified as a world heritage site by UNESCO in 1988, and is located in the prefecture of Macenta in Guinee Forestiere. It is the largest classified forest in the country, covering 119,019 hectares. It is

located at an altitude of 500-1387m. It is part of a 30,000ha forest concession. The forest is currently 27% degraded. It is threatened by hunting. It should be noted that 750 animals a year are taken from this forest (James HEATHER, 2003). It is subdivided into two zones, Ziama 1, with a surface area of 56170 ha, which is subject to multiple uses, and Ziama 2, with a surface area of 42547 ha, which is the central core of the massif that remains today and which must be protected with its zoocenosis.
According to MEDD [8](2009), the Ziama forest is shrinking by 1,111 ha a year, or more than 45,000 m^3 of exploitable wood destroyed each year (40 m^3 /ha/year). It forecasts 5% annual destruction after 20 years.
It seems that in the middle of the XIXeme century, the Ziama forest massif was one of the most densely populated and prosperous agricultural regions of southern Guinea (Benjamin ANDERSON in James FAIRHEAD, 1994) demonstrates this well through his writings during a trip to this region in 1868. According to him, the area was once densely populated, economically bustling, intensively farmed and covered with farms, forest fallows and meadows. He mentions a whole series of villages with populations of between 2,500 and 7,000 people.
It should be noted that a number of constraints are currently hampering the sustainable management of the resources of the Ziama forest massif:
The lack of material and financial resources, the lack of motivation for participatory approaches, the lack of vertical and horizontal communication in state institutions and the inexperience of young curators.
The consequences of demographic pressure, land uncertainty and the population's lack of interest in conserving natural resources.

1-1-1-Research question

How can we ensure the sustainable management of the Ziama classified forest in a context where the concerns of the local population are often at odds with those of the State?

1-1-2-Hypothesis

In the light of our research question, we formulate the following hypothesis: Effective consideration of the concerns, knowledge and know-how of the various stakeholders and their genuine involvement in the regulatory processes are essential conditions for sustainable management of the Ziama classified forest.
Our hypothesis has three dimensions. Firstly, it considers the involvement of local people (as *stakeholders in the management of* the forest) to be an absolute necessity for the sustainable management of the Ziama forest (the forest being

[8] Ministry of the Environment and Sustainable Development

considered a common resource). Secondly, the concerns/needs of all the stakeholders involved in the management and development of this resource must be taken into account. This would appear to be one of the sine qua non conditions for sustainable and balanced management of the Ziama forest massif. These stakeholders include: the State (through its decentralised services, local populations, environmental defence NGOs, etc.). The aim will be to identify the scientific concerns in terms of the ecosystem, biodiversity and climate change and its consequences for production activities, as well as those relating to the living and production conditions of local people.
The second dimension of our hypothesis is *the sustainable management* of the Ziama classified forest in terms of system, method and/or approach. Sustainable forest management guarantees the biological diversity, productivity, regeneration capacity, vitality and ability of the forest to fulfil, now and in the future, the relevant economic, ecological and social functions at local, national and international level, without causing damage to other ecosystems.
The third dimension is the mastery of the rules of the game by all players through the popularisation of regulatory texts relating to the management of forests classified as Gurnee, while ensuring that the provisions of the forestry code are applied.

1-1-3-Objectives

With a level of degradation of 27%, the classified forest of Ziama is today the only "virgin" forest in the country whose protection would be an imperative necessity. In view of the fact that its area and species are still largely undegraded, we have set our objectives as follows:

1-1-3-1- General objective

To contribute to understanding the factors that explain the current level of degradation of the Ziama forest, albeit relatively moderate, and the issues involved in adopting a participatory approach to managing this common resource.

1-1-3-2-Specific objective

The following specific objectives result from this general objective:
To analyse the role of local populations in the management of the massif's resources linked to their livelihood.
To identify and analyse a number of issues relating to the adoption of a participatory approach to the management of Ziama's forest resources with a view to the sustainable development of better governance of this massif.

1-1-4- Research protocol

Having defined the problem, the research question, the hypothesis and the objectives relating to our subject, we now need to define our research protocol.

With this in mind, we will begin by identifying the scientific and social relevance of our chosen subject, then present the reasons for the choice of application site, before outlining our method for collecting, processing and analysing the data.

1-1-4-1- Choice of study area

The Ziama forest is one of the one hundred and sixty-two (162) classified forests of the Gurnee, located in the south-east of the country in the forest region, precisely in the prefecture of Macenta. It covers an area of 119019 hectares, divided between five (5) sub-prefectures, namely: Oremai, Singbedou, Zebela, Fassanconie and Seredou (which has most of its territory invaded by the said forest). It covers twenty-eight (28) villages, including enclaves and riverside communities.

For the sake of accuracy, two enclaves have been chosen: Zoboroma (which belongs to the Oremai RC), and Ballasou from the Seredou RC, then two neighbouring areas: Irie and Avillisou, all from the Seredou RC.

1-1-4-2-Justification of the choice

Our research focuses on a forest classified as humid, called Ziama, located in the south-east of Guinea. This natural forest covers an entire mountain range at an altitude of 1230m. This choice is justified firstly by the importance of the ecological potential of the massif in question from the point of view of both phytocenosis and zooconosis. It is the only remaining forest in Guinea with only 27% degradation compared with all the classified forests in the country. Its protection is therefore imperative. The massif covers twenty-eight (28) villages, but our choice concerns the villages of Ballassou and Zoboroma. These two villages are enclaves, i.e. they are located at the heart of the massif. Most of the socio-economic activities of their populations involve exploiting the resources of this forest, with only user rights to certain plots, which are insufficient for their subsistence. Avillisou and Irie are riverside villages. But their populations had exploited the forest under the first regime by growing export crops. Disenfranchised in 1991, these village communities are demanding their return to the areas they once exploited.

These twenty-eight riverside villages present the same situations, but the four targets were used as a sample because of their representative nature, their two main characteristics: on the one hand, enclaves, and on the other, riverside dwellers.

1-2- Methodological approach

In order to gain a better understanding of the problems associated with the management of Ziama's forest resources, we are using approaches that have been tried and tested in the forestry sector in tropical countries, such as

participatory or community-based, eco-systemic and integrated approaches.
Of these approaches, we feel that the participatory or community-based approach is the most appropriate for this study because it takes into account the needs of the population and involves all the stakeholders.
This study offers detailed portraits of the modes of governance of the Ziama forest ecosystems. We share an approach that combines recent social theory with an ability to describe intimate details of the transformation of institutions for the management of this forest. In addition, we seek to reveal and analyse some of the consequences and possibilities of a genuine revolution in forest use and management relationships. Whether it be the relationship between individuals and nature, between administrators and communities, or between communities and communities as evolving national actors. Human beings negotiate their relationship with nature, but this relationship differs from one environment to another and from one era to another.
To carry out this research exercise, we opted for a methodology combining qualitative and quantitative approaches. Interviews and documentary research were the main data collection techniques used. The surveys focused on three (3) target groups: government technical services, NGOs and local residents.
For the State's technical services, our surveys mainly concerned the Forestry Centre and the forestry cantonment, which are the services responsible for managing and regulating classified forests in Guinea. The size of the sample at the level of the State services was four (4) individuals with many years and various experiences in the field: the head of the protected area management division at the National Directorate of Water and Forests, the head of the Ziama branch, the head of the N'Zerekore conservation division, and the head of the Seredou forest cantonment.
In the riverside villages, we carried out focus groups with three population strata: young people, women and men. We interviewed thirty-two (32) men, seventeen (17) women and twenty-four (24) young people. This number covered all the villages surveyed. However, the size of each group varied from village to village. During the interviews, each person wrote down the details of their answers in a notepad. Some interviews were recorded using a drinking telephone. At the end of each day's interviews, we met again to take stock, complete the notes taken and transcribe the recordings. A summary report was drawn up at the end of the survey in the CR of Seredou in the presence of the services responsible for managing the massif and local elected representatives. The final transcription was made in Conakry, which enabled us to draw up the second report.

Table 1: Age of respondents by sex and by village

Villages	Men	Women	Young people	Age range	Total per village
Ballassou	9	5	8	28- 80	22
Zoboroma	10	6	7	30 - 70	23
Avilissou	7	3	5	29- 100	15
Irie	6	3	4	30 - 80	13
Total workforce	32	17	24		73

Source: Candidate survey (August 2014)

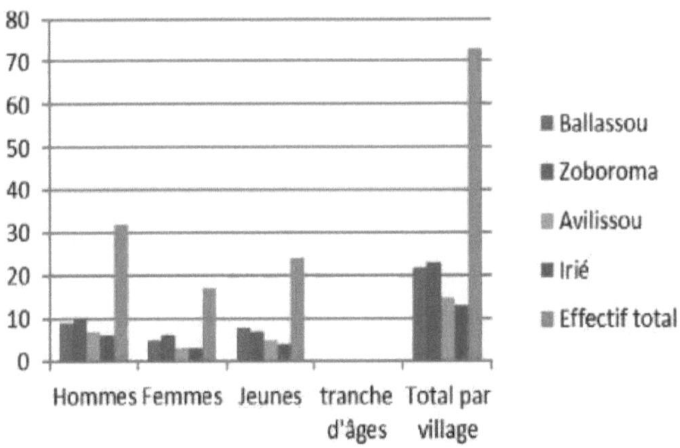

1-2-1- Documentary research and analysis of secondary data

As part of this research, the literature review consisted of consulting available documents relating to our subject.

In order to obtain a wealth of information on the subject, we have drawn on the bibliographical data needed to understand the current situation. This literature is admittedly exhaustive, but given the abundance of literature on the issues involved in the management of common property forest resources, we felt that it would be useful to take a closer look at it. To avoid getting bogged down in the exploration of our field of study, we made a bibliographical selection. To do this, we benefited from the essential contribution of some of our professors from the Master's degree in Space - Time - Society, who provided us with the reference documents and recommendations necessary for our subject.

This phase mainly involved general literature, scientific journals, articles, study reports, dissertations and any other document related to our research topic, as well as websites.

The results of the literature review enabled us to give a degree of precision and clarity to the original question. They also helped us to situate the problem.

This literature search enabled us to explore previous research carried out elsewhere in the world on forest resource management, as well as the theoretical approaches developed by various authors.

These results also enabled us to enrich the problem and the conceptual and theoretical framework, on the one hand, and to gain knowledge of forest management methods in Gurnee and Ziama, on the other.

1-2-2- Collecting primary data

The aim of collecting primary data is to clarify, correct or improve the level of information provided by current trends relating to the problematic of the subject on the basis of literature reviews.

Depending on the methods chosen, we used two main tools: the questionnaire, which included closed and semi-closed questions. It enabled us to assess specific quantifiable parameters in order to gain a better understanding of the following aspects:

- Respect for the laws governing forest regulation by local people;
- Community participation in drafting forestry legislation;
- NGO support for the technical services in charge of managing this forest;
- The difficulties faced by the local population.

The second tool, the guide, was based on individual and group interviews.

Using the interview guide, we collected the information we needed to deepen our analysis using a qualitative approach, in particular the following aspects:

- History of the founding and settlement of the villages ;
- Rules governing the regulation of the forest prior to classification;
- The consequences if someone breaks the rules;
- The importance of the forest for local people ;
- Activities carried out in the past by the local population;
- The relationship between the aunts and the forest;
- The date on which the forest was classified;
- The difficulties faced by local people from the classification of the forest to the present day;
- Internal community problems;
- The nature of the conflicts that arise between stakeholders in the forest ;
- Dispute resolution methods ;
- The resurgence of conflicts and the dates on which they broke out;
- Ownership of this forest today.

During the interviews, we kept a note of the responses. Some of the interviews were recorded using a drinking telephone. At the end of each interview day, we got together to take stock, complete the notes taken and transcribe the

recordings. A summary report was drawn up at the end of the survey in the CR of Seredou in the presence of the departments responsible for managing the massif and local elected representatives. The final transcription was made in Conakry, which enabled us to produce the second report.
Data was collected in the field from 18 August to 05 September 2014, i.e. for three weeks.
Based on direct observations of the lowlands, which are the main cultivation areas in this zone, we have estimated their surface area at 1 ha per person. These areas vary according to needs and villages. Distribution among farmers is entrusted to a management committee set up in each village. The age of our respondents was calculated on the basis of their identity cards.

1-2-3-Sampling

Whether a survey is conducted by questionnaire or by interviewing individuals or groups, sampling is always necessary to ensure the representativeness of the sample. Because of our limited material and financial resources and the remoteness of the area concerned by our subject, we collected statistics on the various conflicts from one hundred (73) heads of household spread over four villages surveyed. This was because these villages all had similar situations.
As mentioned above, this forest covers five (5) RCs, spread over twenty-eight (28) villages. However, we have chosen certain RCs and villages; this choice was justified by their demographic size, which is considered here to be the main factor in the changes that have taken place, both economically and socially and culturally. This is the case, for example, of Irie, Ballasou and Avillisou (in the RC of Seredou) and Zoboroma (in the RC of Oremai).
Our interest in the choice of the two enclave villages lies in the fact that their population seems to be forced to adapt to the narrowness of their village lands because they are experiencing strong demographic growth.

Table 5: Demographic size of villages surveyed

Villages	Residents
Irie village	3362
Irie crossroads	1939
Ballassou	509
Avilissou	3428
Zoboroma	3500

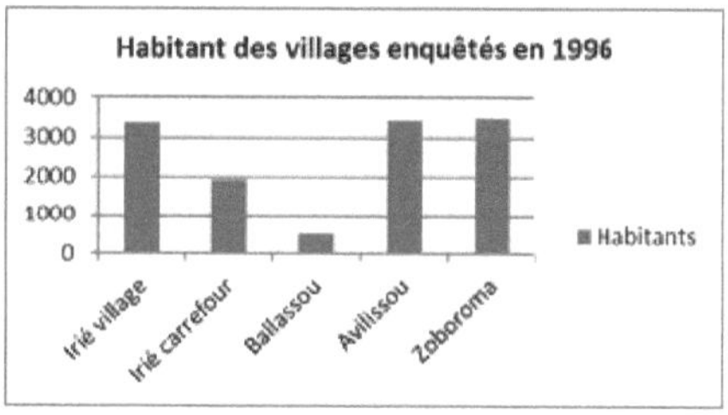

Source : CR de Seredou

In the past, it was the tribal chief who managed the land on behalf of the community. He distributed it and the area cultivated depended on the number of family members and the type of tools they owned. There was no individual ownership. The abundance of land corresponded to the cultivation system in place, which was none other than itinerant. Land was a resource that was developed and then left. The land tenure systems were adapted to a society in stationary equilibrium, based on traditional anti-individualist law, aimed above all at ensuring the security of the group or of the individual within the group or by means of the individual within the group.

With a small population, they were able to grow crops long enough to replenish the soil.

With the demographic growth observed in the riverside villages. The first consequence is the imperative need to increase food crops. Food requirements have increased, and if farming methods are not rapidly improved, unsatisfactory agricultural production could threaten the already precarious ecological balance. The abundance of land is no longer a permanent feature, but farmland is in short supply in some villages such as Zoboroma and Ballasou.

The increase in population meant that land use had to be reorganised, either by consolidation, because the plots had become too small.

Toma society was egalitarian when it came to land use. Everyone had access to the land according to their needs. The alienation of cultivation plots (lowlands) to non-Toma by the forest guards of the time destroyed this egalitarian concept. This inequality has only increased, and now some families have reached the point where they have no arable land at all.

1-2-4-Conceptual and theoretical tools -Analysis method

We begin this section with a critical presentation of the theories that have been used in forest resource management, and the definition and evolution of certain concepts that are the focus of our analysis. We regard these concepts as operating tools for data collection and analysis in the context of our research.

This conceptualisation exercise also makes it easier to understand the dimensions of the problem relating to our research object. It is therefore an indispensable aspect of the design of effective tools for collecting and analysing information in the field.

These concepts include: the concept of actors, the concept of governance and the concept of resources. The understanding of all these concepts was facilitated by the mobilisation and exploitation of a large body of literature on the subject. There are three theories: the ecosystem approach, the integrated approach and the community or participatory approach. But of these three, the one that has caught our attention is the participatory or community approach.

1-2-4-1-Evolution of participatory or community theory

The participatory dimension of local populations seems unavoidable today when tackling the issue of forest management (Maya LE ROY, 2013). However, it has long suffered from a deep-seated ideological rivalry between the 'conservationist' school on the one hand and the 'communitarian' school on the other.

Conservationists believed that the population did not have the right skills to manage the forests. In their view, forests should be conserved to the exclusion of all human activity. We note with Rebeca HARDIN (1968), who thought that entrusting the management of a common good such as a forest to the communities would be synonymous with massive destruction. For her, it would be better to '*privatise*' or '*etatise*' the management of natural resources. The argument of the school of the tragedy of the commons is that in the presence of a common good, individuals who think only of maximising their profit will end up depleting the resource.

As for the 'communitarians', they campaigned for the use of forests to be taken into account by the people who depended on them. For the proponents of this school of thought, such as Elinor OSTRON, neither the State nor the market has the power to manage forests. In their view, societies have long since developed their own systems for managing their resources. A system based on rules that have enabled them to live in harmony with nature, expropriating this population from the resource they are used to managing would not be a reliable solution for the sustainable management of natural resources. So we need to give them a sense of responsibility, and involve them in some way in their management. Elinor OSTRON thus advocates a decentralised mode of governance.

As in Ziama, where conventional state management has contributed to the destruction of the massif, increasing the pressure on vegetation outside village areas and upsetting local people's opposition to the idea of conservation.

A third group, the "loggers", oppose the communitarians and support the conservationists. For them, local populations are an obstacle to logging activities. They must therefore be excluded.

In 1970, communitarian thinking gained momentum, driven by certain NGOs that were beginning to defend the position of indigenous and local communities in a highly publicised way. A case in point was the media frenzy in 1989 surrounding the international campaign by Chief Raoni Metuktire of the Koyapos people to defend Amazonian tribes and combat tropical deforestation. According to this position, *"the appropriation by international bodies of a line of thought favouring participation and management by communities marked a profound break with the traditional, highly centralised schemes, which tended to promote the importance of the role of the State in the management of renewable resources"* (Maya LE ROY, ibid, p 23).

The year 1990 marked the beginning of decentralisation with the advent of the Structural Adjustment Programme in the countries of the South by the IMF and the World Bank. This programme gave weight to participatory approaches. As a result, participatory projects became widespread. Pascal Rey (2007, p 252) points out that "the *participatory* approach is making *a comeback among all development players, from major international institutions to small NGOs*". This has put forests firmly on the international agenda.

In fact, although the participatory or community forestry approach has been the subject of much discussion between the different schools of thought on sustainable forest management, it seems to be appropriate to the problem of this research.

1-2-4-1-1-The concept of actors

Since the forest is a common resource, its management necessarily involves the participation of several stakeholders at various levels. These stakeholders include the state, NGOs, local authorities and grassroots communities. Each of these actors must, in one way or another, be involved in maintaining the achievements of nature in general and, more specifically, that of the forest, which is both a source of income in tropical areas and the subject of conflict between populations and between populations and the forestry administration. This brings us to the concept of the stakeholder in this study.

This concept appears from various angles (sociology, geography, etc.) and its meaning depends on the current to which it belongs. We will therefore use it from a geographical perspective.

In the social science literature of the French language, the intensive use and plasticity of the concept of the actor, its metaphorical value and its ambivalence, allow a number of bridges between disciplines, but it also contains ambiguities. The term "actor" has become a "password" in the social sciences, although it is

not necessarily reserved for those who know its precise meaning (Jean Pierre GAUDIN, 2002.in http :/ress.revues.org).
In common usage, the reference to the social actor suggests above all freedom of initiative, but also the resources of rational behaviour and, in short, is associated with the figure of the modern individual.
In the social and human sciences, constructions of the notion of actor are very diverse, and its uses are multiform, even at first analysis. Philosophy, for its part, has essentially relied on the subject/object distinction. In law, as in psychology, the notion of actor is similar in its uses to those of subject or rational individual.
The initial versions of organisational sociology from the 1950s talked more about the problem of rational choice. As for Marxist approaches, which are more structural in inspiration, they tend to talk about agents in contexts or constraints that explain their actions (macroeconomics, structuralist sociologies, etc.).
Over and above conventional disciplinary knowledge, the concepts of 'actor' and 'agent' have been received in very different ways in the social sciences.
The concept of "agent" was first used by Weber MAX in 1921. Etymologically, an agent is someone who acts, causes, is at the origin of [...]; used more in economics than in sociology. It was thus used by sociologists, retaining its originality until 1980.
In the 1970s, the notion of "agent" changed to the concept of "actor" under the impetus of Michel CROZIER and Erhard FRIEBERG [9](1977), for individuals or groups of people with the capacity to change people's behaviour. They developed the theory of strategic actors. This theory became central to the sociology of organisations, and was particularly developed within strategic analysis.
For them, the notion of "agent" was used for the one who transmits; the vector of transmission (idea or disease). According to their analysis, the word had a weak meaning, meaning a simple intermediary, i.e. a representation or relay.
The concept of actor has largely supplanted that of agent, which was the term originally used and which was dominant until the 1980s.
The return of the concept of the actor is justified by four factors: the rejection of determinism, the return to micro-sociological approaches, the need to analyse social change and the desire to take account of conflict.
All these concerns can be found in the objects of sociology, but the question of

[9]We have drawn this analysis from the writings of Michel CROZIER and Erhard FRIEDBERG, 1977 "l'acteur et le systeme : les contraintes de l'action"; they are all thinkers in the field of organisational sociology.

the actor has been at the heart of three programmes in particular: the problem of organisation, rational choice and the sociology of action.

The use of the notion of actors is tending to become legitimate in Geography, but it seems to be a Sociological invention. Indeed, as sociology claims authorship of the concept, we are reminded of Simon OSTROWETSKY's rule (1997) in the following terms: "*If we want to think in terms of a meta-discipline, we must not go looking for definitions belonging to the fields of sociology from people who are not sociologists. You have to know which sociologist you are talking to*".

Geographers use the concept "actor", which is always accompanied by : "Territoire". It was first used in geography by Pierre GEORGES in 1964 and by Jean LABASSE in 1966. "*Territories are subject to the intervention of multiple institutions, social groupings, political forces and individuals of varying degrees of power, depending on their purpose, which are assigned by these actors. They act by virtue of ideological projections and decisions supported by private and/or public investments of varying degrees operating at different spatial levels. They participate in the general and partial dynamics of territorial transformation and adaptation*" (Dictionnaire Geographique, 2005, p. 6-7). Sociologists, for their part, put forward the concept of actors.

In this debate on the concept of actor, geographers give priority to the notion of 'spatial dimension' and 'territory'.

The sociology of organisations, for its part, divides this concept into "*systemic and strategic analysis*", or "*system of concrete action*", and "*system of organised collective action*" (CERTU, 2001, p8).

Michel CROZIER (1977) classifies actors into two categories: private actors (international institutions and NGOs) and public actors (the state, local authorities and farmers' organisations). For him, the major elements of a territory are its private or public players. Michel CROZIER also points out that the boundaries between an organisation and its environment are not watertight. He believes that there should not be "organised organisations on the one hand, and unorganised fields on the other". It is often impossible to draw the boundaries, and we have to agree, in his view, that "there is no field that is not organised in terms of power".

Erhard FRIEDBERG, Michel CROZIER's ally, talks instead of "*systems of collective action*".

the organisation. For him, an organisation's environment is made up of other organisations.

For geographers such as (Guyi DI MEO, 1996 in http://eegeosociele.free.frirenne, 2006) "we need to take account of what the act [is] about". However, the sociological context of the concept of actors leads us to overlook certain forms of the geographical relationship. In social geography, it is part of a logic of redistribution of epithets (subject, individual, person, actor) often used to analyse geographical spaces.

According to Jean Pierre VANDENBERG (1998), stakeholders can be divided into three distinct categories (micro-stakeholders, macro-stakeholders and mega-stakeholders), distinguished by their roles.

The former take action one step at a time and have little impact on the social structure. As for the last two (i.e. the macros and the mega-actors), their actions can have quite considerable social ramifications, either because they speak and act on behalf of the social actors, or because although they speak in their own name, "they have significant power".

According to Richard GRIMBLE, et al, (1997, in www.icra-edu.org), there are "active" players, "passive" players, "key" players, "secondary" players and "primary" players.

But they are distinguished from each other by the role they play in the system. Active" players are those who have an effect or determine a decision in the system or project, while "passive" players are people who are affected by the decisions or actions of others.

The "primary" stakeholders are the intended beneficiaries of the projects; the "secondary" stakeholders are those who act as intermediaries within the projects. As for "key" players, these are people who are considered important or who have a significant influence on the success of a project.

A great deal of literature has been written on the classification of stakeholders, but based on the evaluation criterion, we classify stakeholders into two types: 'important' stakeholders and 'influential' stakeholders. Important" stakeholders are people whose needs are important to a project or study. Influential stakeholders have the power to control activities or can influence the decision-making process.

We are taking women farmers as important players here, but if we want to judge equality, traditionally they will not have any influence in the decision-making process.

Socio-anthropologists refer the use of the notion of actors to the context of the action, while clearly setting out the limits of the analysis thus carried out or to be carried out.

According to Hubert (1992), a thinker in the structural geography movement, "*the notion of social actors is a model for qualifying life trajectories that allows us to escape any concrete, pre-constructed spatial reference*".

Analysis of the concept of actor opens up a wide field of discussion. Whereas

social geography aims to maintain the originality of terms in its analyses, in both explanation and description, which would lead to the question of localisation.
It should be noted that the notion of 'space' remains at the heart of the disciplinary core of geography.
Based on our analysis, we propose the following definition of the concept of actors.
"A stakeholder is a person or a group of natural or legal persons who intervene in an area to solve a problem". For example, the NGO Danish Refugee Council[10] , the farmers' organisation for aid to elephant victims, local residents and local populations are all stakeholders in our research project.
As part of the sustainable management of forest resources, the players involved must work together and, above all, be aware of the concerns of local people.
In the specific case of Ziama, which is the subject of our research, there is a problem of neglect of essential needs and a lack of understanding of the role of the local population. The main, and indeed essential, stakeholder, the local population, is not taken into account in decision-making bodies. The exclusion of the local population from decision-making is not confined to the Ziama region, but is a general phenomenon throughout the country. Yet sustainable development is based on a strong idea of the participation of all stakeholders.
The texts relating to the management of forest resources dictated by the competent administration are not applied by the population, because their contents are not certainly popularised and/or do not take into account the needs of these populations. This is creating real conflict in the area.
The analysis of actors concerns groups such as communities, government or the private sector, which are assumed to be completely homogeneous. But these groups are clearly not homogeneous.

1-2-4-1-2-The concept of governance

The general concept of governance applies to several fields and its meaning depends on the disciplinary specificity of the person using it. In this particular case, we will attempt to clarify it through its etymology, its history and its use as a conceptual tool in our research.
The meaning of the word 'governance' varies widely. The term is often used today in both English and French, without the meaning being clear. Charlotte Gisele Kouna Eloudou (2012) describes governance as a 'vague, ambiguous,

[10] We met this NGO in Avillissou on 22 August 2014 during our survey, which is involved in resolving the conflict between the neighbouring villages of Avillissou and Koima. We also met another organisation, a farmer's organisation, in Irie, in the same context, a little different from the Danish one, which is a community initiative set up to help elephant victims, i.e. those whose fields have been ravaged by elephants.

catch-all concept, with ideological connotations and poor analytical skills, which has emerged in a particular international context'. It is seen as "the latest in a series of state reforms proposed to the countries of the South". For Moreau Defarges (2011), governance, like decentralisation, has a variety of meanings and lends itself to many uses. These meanings and uses are sometimes contradictory and have evolved over time, depending on the context (political, economic, cultural), the user and the sector of activity. INRETS [11] (2009), talks more about the institutional dimension of the concept. According to this institution, governance is often presented as an integral part of the concept of sustainable development. But it was not born with it.

1-2-4-1-2-1- The emerging context of governance

The concept of governance derives from the Greek verb "kubernan", which means "to steer a ship or a chariot". It was first used metaphorically by Plato to describe the act of governing people (INRETS, 2009). It gave rise to the Latin verb *gubernare,* which had the same meaning and which, through its derivatives, including *gubernantia,* has itself given rise to numerous terms in several languages:

In Old French, "gouvemance" was first used in the 13th century as the equivalent of "gouvernement" (the art or manner of governing) and then, from 1478, to designate territories with a special administrative status. It then entered the English language in the 14th century, giving rise to the term *governance;* in Portuguese *governance;* and in Spanish *gobernanza.* It is said to have been used mainly to mean "government", but with the same meaning as in French (Jean Pierre GAUSSIN, 2002).

All these terms subsequently fell into disuse, in some cases (France, Portugal) because they were associated with the ancien regime.

The English word *governance* was revived in the late 1930s in the corporate context. The proponents of local participatory democracy stemming from the urban social movements and self-management ideologies of the 1960s and 1970s then seem to have been the first to reuse the notion of governance.

In a third phase, the technicians involved in modernising public management came from a different ideological background, followed from 1989 onwards by the major development aid institutions, especially the World Bank, which were the decisive agents in popularising the term. *Governance* was then adopted by academic and mundane analysts of globalisation and international regimes, and then by the European Commission, in the form of a truly constructed concept .[12]

[11] Institut National de Recherche sur les Transports et leur Securite.

[12] Our reflections on the history and etymology of the concept of governance are essentially inspired by the work cited in INRET, (2009), where a number of authors have been consulted.

In the 1980s and 1990s, the notion of governance gained renewed interest and slipped into the political and development fields. It also made inroads into the environmental field in the 1990s. This renewed interest is linked to a number of factors: the financial crises and failures of states to manage collective problems, the decline of state sovereignty and traditional government, and the 'governability' difficulties of modern societies (Charlotte Gisele Kouna Eloudou, 2012). Other factors have contributed to this renewed interest, including the mixed results of globalisation and the restructuring of decision-making levels in the context of globalisation (ibid.). Moreover, this emergence has been fostered by the institutionalisation of the market as a regulatory space, globalisation and the rise of non-state actors. The strong criticism of multilateral institutions by civil society, the emergence of a new management of collective action and the difficulties of 'political steering' in a multi-actor context are also factors that justify the renewed interest in governance.

The effective participation of all stakeholders is one of the essential pillars for initiating sustainable development. The aim is to ensure that all stakeholders are involved in the management of natural resources in general and forestry resources in particular. This would enable grassroots communities to take control of their own resources and give everyone the opportunity to express themselves freely within the framework of good forest governance. The latter is defined as a set of rules and reinforcement mechanisms, as well as interactive processes that regulate the activities of stakeholders towards a concerted end. It is also seen as the establishment of shared management processes, regimes and structures, both informal and formal, that facilitate coordination, concern and complementarity between governments and organisations in the same field.

Some believe that good governance is undoubtedly the most important factor in eliminating poverty and promoting sustainable development. In any case, it enables stakeholders at all levels to be involved in the management of forest resources.

But the process is not without its difficulties, as each stakeholder seeks to pull the wool over its own eyes. So, to overcome the shortcomings and impacts of local forest governance and ensure sustainable forest management, it is important to situate forest management issues within a more global framework of governance. Forest governance has been made possible thanks to the pressure exerted by local populations and civil society on governments and the international community, which have felt the need to promote participatory sustainable management of forest resources by taking major political, legislative and regulatory measures to encourage the effective empowerment of populations. Today, this political, legislative and regulatory arsenal forms the basis for a change of vision and strategic approaches aimed at promoting sustainable management of forest resources and involving local populations and civil society organisations. (*synapse.uqac.ca/wp.../**Forets_and_humans_Complete_Study_Chap_05**.pdf*)

According to Abdelilah Baguard et al, (2012, p. 4) *"in forest management, stakeholder participation in decision-making processes is recognised as an approach that can ensure the sustainable management of forest ecosystems, taking into account their multifunctionality".*

In the aftermath of the Rio de Janeiro conference, all the signatory nations expressed their determination to achieve good governance of their forest resources. This presupposed awareness and the responsible participation of all players in the search for rational solutions to the problems of economic, social and cultural development.

Cardinal principles have been laid down, including the decentralisation of the decision-making process and the empowerment of the population through an inclusive participatory approach, to lay solid foundations for sustainable human development worldwide.

From the Rio de Janeiro conference in 1992, through Johanesburg in 2002 and Kyoto in 2012, not forgetting the Nagoya conference in 2010 on the Convention on Biological Diversity, many relevant decisions have been taken, but their application in some countries is lacking. In West Africa, for example, vegetation cover is estimated at 72 million hectares, or 14% of the total land area. However, the rate of deforestation is higher than the global average (FAO, 2009).

1-2-4-1-2-2- EVOLUTION OF ITS USE

The challenges of sustainable natural resource management make it imperative to change legislation and practices in most African countries. In West Africa, the authoritarian, centralised management system for land and natural resources inherited from colonisation is increasingly giving way to concerted, decentralised management mechanisms (Adam Malam KANDINE, 2008). In other words, the wind of political democratisation that blew across the African continent in the 1990s led public authorities to review the main public policies, including those relating to land tenure and natural resource management. With regard to the management of natural resources, it should be noted that the international context has also evolved considerably, notably with the adoption of a number of conventions which have established founding principles that oblige signatory states to change their policies in this area. According to Adam Malam KANDINE, the concept of governance in Africa was born or reactivated at the same time as the concept of participatory management, i.e. in 1990. In fact, transparency and good governance are seen in Africa as two fundamental requirements laid down by the International Monetary Fund (IMF) and the World Bank as part of the implementation of the famous Structural Adjustment Programmes (SAPs) to which most sub-Saharan African countries were subjected from the 1980s onwards. This exhortation to better governance coincides with the opening up of democracy, which not only imposes pluralist electoral competition but also authorises the participation of several categories of non-state actors in the management of public affairs [13]. Consequently, the concept of (good) governance could not find a more favourable terrain for its implementation. Catherine Baron (2003) emphasises the functional aspect of the concept of governance in Africa. She defines it as "the art or manner of governing by promoting an original way of managing affairs in an environment characterised by a plurality of players (a firm, a State, a local authority, a non-governmental organisation, an association or an international body), each of which has decision-making power to varying degrees and in a more or less formal manner" 1 [14]. The implementation of this concept has also led supporters of its promotion to give it sectoral declinations. For the International Union for Conservation of Nature (IUCN), governance of natural resources means the interactions between the structures, mechanisms and traditions that determine how power and responsibility are exercised, the decisions that are taken, and the individuals and other stakeholders who are in a position to give their opinion on the management of natural resources - and in particular the conservation of biodiversity .[15]

According to studies by Adam Malam KANDINE14 , the aim of the concept of governance is to assess, positively or negatively, the way in which public affairs are conducted. A certain number of characteristics are used to measure the actions of the various players, namely: defining a clear direction, transparency and real results. But they must also be based on fundamental human rights and values, in particular honesty and fairness, as well as real commitment and contribution to decision-making. The FAO [16] [17], on the other hand, talks more

[13] See, Koffi Allinon, societe civile togolaise face aux défis de transparence, de bonne gouvernance et d'efficacite : etat des lieux et perspectives, in Question de gouvernance, les cahiers de Mande Bukari, revue trimestrielle de l'universite Mande Bukari, N°04, 3eme trimestre 2006, P.4-13.

[14] Catherine Baron, "La gouvernance : débat autour d'un concept polysemique", in Droit et Societe, 54/2003, P.330.

[15] IUCN, World Conservation Congress, 3rd Session, Bangkok, Thailand, 17-24 November 2004.

[16] Our analysis is based on the study carried out by Adam Malam KANDINE in 2008 in Niger on Land and Natural Resource Governance: The State of Play in West Africa. In this study, he emphasises the cardinal principles that characterise governance in general and in African countries in particular.

[17] See the study carried out by the FAO in 2007 on "Good governance of land tenure and

about accountability, political stability, effectiveness on the ground, regulation and the effective application of laws. Thus, the concrete implementation of the principles of governance requires fairness, efficiency, transparency accompanied by accountability, sustainability, civic commitment and security.

1-2-4-1-2-3-A concept characterised by its polysemy

In its contemporary version, governance is understood in a broader sense than that of "government". It is the subject of numerous definitions for which no consensus has yet been reached. Georges PAQUET (2008) defines it as "effective coordination when power, resources and information are widely distributed among several hands". For him, governance "is at once a way of seeing things, a diagnostic tool and an instrument of clinical intervention: it helps to detect and pinpoint the nature of the flaws, to understand where the breakdowns come from, and to develop appropriate corrective measures". In this vision, which refers to power and communication, governance is a "one-stop", "multi-functional" concept which is at once a method of observation, a tool/instrument for identification and action, and a solution. Thus, Georges PAQUET sees the effectiveness of governance as being conditional on the combination of efforts through the mobilisation of financial and human resources. However, the mode of governance put in place at Ziama suffers from inefficiency.

According to the Commission on Global Governance (CGG, 1995), governance refers to "all the different ways in which individuals and public and private institutions manage their common affairs. It is a continuous process of cooperation and accommodation between diverse and conflicting interests. It includes formal institutions and regimes endowed with executory powers as well as informal arrangements on which people and institutions have agreed or which they perceive to be in their interest".

In analysing this definition, she identifies three main elements: the actors and their interactions, the consensus, and the mechanisms for regulating social processes and decision-making. It can be criticised in two ways. It refers only to public and private players. However, in the context of (local) development, there is a diversity of actors who cannot be classified solely in these two categories. Moreover, these actors have different resources and social status, and are involved in the creation and management of goods and in the provision of public or collective services. Moreover, in the second part of this definition, it is consensus that seems to determine the existence or absence of governance. In other words, there is no governance or there is bad governance without

---

administration".

consensus. However, the fact that all societies produce decisions, standards and goods and deliver services, at any historical moment and under all the political regimes in place, shows that governance exists in all societies (Olivier DE SARDAN, 2007).

According to Isabelle LACROIX et al (2012), in her book 'Gouvernance': tenter une definition, defining a concept as broad and uniform as governance is a definite challenge. For her, this term is used in "all sorts of ways", so there is a real need for clarification in the literature . [18]To do this, we use the works and contributions of authors in different fields and disciplines.

Many authors mention the indefinite aspect of the concept of governance.

For Jean Pierre Gaussin (2001), governance refers more to the modification of the relationship between politics and economics, whereas for Bernard Jouve (2006), it refers more to the questioning of the governability of traditional Western democratic societies run by a single central decision-making authority.

According to the World Bank, governance is the set of traditions and institutions by which power is exercised in a country with the objective of the common good of all [...].[19]

We find this definition interesting in that it links the exercise of power to the search for, identification, classification and management of the common good.

The European Commission, for its part, offers a definition adapted to the context of European development, but which, in our view, offers a certain potential for generalisation of its use. For this institution, "*the notion of governance refers to the regimes, processes and behaviours that influence the exercise of power at European level, particularly from the point of view of openness, participation, accountability, effectiveness and coherence*".[20]

The UN [21], through UNDP , [22]also offers a definition of governance that is likely to be in line with international realities. For this UN body, "*governance can be seen as the exercise of economic, political and administrative authority to manage a country's affairs at all levels. It comprises the mechanisms, processes and institutions through which citizens and groups articulate their interests, exercise their legal rights, meet their obligations and mediate their differences*".[23]

---

[18] Isabelle Lacroix and Pier- Olivier St-Amaud.
Professor of applied politics at the Université de Sherbrook.

[19] This definition is taken from a World Bank report entitled "Collaborative Governance" at http: //www.worldbank. org/wbi/governance/eng/about-e.html#approach). Accessed 04 April 2014.

[20]http://eurlex.europa.eu/LexUriServ/site/fr/com/2001/com2001 0428en01.pdf

[21] The United Nations.

[22] United Nations Development Programme.

[23](http://mirror.undp.org/ magnet/policy/chapter1.htm#b).

This definition evokes the legal rules governing this approach, while referring to the same logic of participation and empowerment. But a new concept is emerging at this level: the notion of managing disputes by means of some form of mediation.
CIDA [24]proposes a more relevant definition that encompasses different notions, focusing on certain components of governance in the following terms, *"governance encompasses the values, regimes, institutions and processes by which individuals and organizations attempt to achieve common objectives, make decisions, establish authority, legitimacy and exercise power".*
(http://www.acdicida.gc.ca/INET/IMAGES.NSF/vLUImages/Evaluations2/$fL1 e /Review of Governance Programming in CIDA-EN.pdf)
This definition seems relevant to us in that it emphasises the notion of value, decision-making, legitimacy and authority, which are in a way the guarantors of power. In the Ziama area, however, there is a total lack of legitimacy and authority, as the massif's resources are suffering from human aggression.
According to the economist Alain Beitone (2007, p 252), [governance is] "*the set of transactions by which collective rules are drawn up, decided, legitimised, implemented and controlled*".
According to this definition, governance is an active process that takes the form of multi-player actions guiding decisions and ultimately actions.
From a political science perspective, [governance] "*refers to all the institutional procedures, power relationships and public or private management methods, both formal and informal, which govern real political action*" (Guy HERMET, 1998, p114). In his definition, Guy HERMET adds to the concept of governance the notions of formal and informal actions, and these two levels must be considered as an integral part of governance.
Philippe Moreau DEFARGES (2003), for his part, offers an approach that defines this concept in terms of the context in which it was born.
In his view, governance, like globalisation, is a concept of the 1990s. This "*notion of governance marks the emergence of new ways of administering societies and international relations. Governance implies permanent negotiations, on an equal footing, between the major players in the system: States, organisations, companies, etc. With governance, the social field becomes a playground. Instead of dictating priorities from on high, the government or authority simply regulates and arbitrates. The aim of the community or society is no longer some great transcendent design, but the free development of the activities of all*" (p. 96).
According to Patrick TRIPLET (2009), governance is "a process of coordinating

[24] Canadian International Development Agency.

actors, social groups and institutions with a view to achieving objectives that are defined and discussed collectively".

Isabelle LACROIX et al, (2012), give a definition inspired by those of the institutions mentioned above. They bring together the key elements of governance embodied in the concepts of rules, processes, interests, players, power, participation, negotiation, decision and implementation. They are thus defined as follows.

"Governance is the set of collective regimes and processes, whether formalised or not, by which the stakeholders concerned participate in the decision-making and implementation of public actions. These regimes and processes, like the decisions that result from them, are the result of consian negotiation between the multiple actors involved. As well as guiding decisions and actions, this negotiation facilitates the sharing of responsibility between all the players involved, each of whom possesses a certain form of power" (Isabelle LACROIX et al, 2012, p 26).

Lamine Mandiang, (2008, p 2), for his part, defines governance as follows: "*governance refers to a way of approaching the question of government which does not give ownership to the art of governing and the techniques for carrying out action, but to the relations between the leaders and the led, in particular civil society and the State*".

From its origins, governance has two main fields of application: corporate governance and political governance.

Corporate governance is the set of processes, regulations, laws and institutions that influence the way a company is managed, administered and controlled.

Governance also includes the relationships between the many players involved (the *stakeholders*) and the objectives that govern the company. The main players are shareholders, management and the board of directors.

Corporate governance enables the company's stakeholders and partners "to have their interests respected and their voices heard in the running of the company".

As well as political governance, in which societies are governed by free democracy, it refers to *interactions between the state and society, in other* words to coalition systems of public and private actors.

It postulates the introduction of new, more flexible modes of regulation, based on partnerships between public and private players. The aim of this process of coordinating different players is to "make *public action more effective and societies easier to govern" (*Lamine MANDIANG, 2008, p 3-4).

Despite the diversity of applications of the concept of governance, there is a common dynamic in its use. Thus, for most of those who use it (private and public sectors), it denotes a movement towards 'decentralisation' of decision-making, with a multiplication of the places and players involved in this decision (Lamine MANDIANG, 2008). The author continues his analysis by saying that decentralisation of decision-making is the sharing or transfer of the seat of decision-making between several centres and players. This is what the riverside communities in our study area are demanding. According to them, their point of view will at least be considered during decision-making and can be influential.

Based on our analysis, we attempt to define the concept of governance in relation to our study as follows: "*Governance is the set of rules and principles that make it possible to manage an issue or a place"*. From this point of view, it must involve most of the players: civil society (associations, citizens, etc.), private sector companies and the various public authorities. The aim of governance is "*to improve the management and effectiveness of an issue or an area through the participation of the parties concerned" (*dictionnaire geographique 2005, p 183).

On the basis of the above, we believe that if the management of the resources of the Ziama massif is to be effective, all the cultural values of this community must be taken into account.

1-2-4-1-3- The concept of resources

Whatever the types of resources mentioned, we give priority to the material resources that are relevant to our research.

Pinpointing the concept of resources remains a tricky, even illusory exercise, as it is one of the most ambiguous concepts in geography. Everything seems to be a resource or could potentially become one.

By way of retrospective analysis, geographers and other scientists at the beginning of the XIXe century gave the study of resources an important place in the development of a process of knowledge (observation, discovery, classification, enumeration). It was at this time that the notion of natural resources became more apparent, with the start of the industrial revolution.

At the end of the XIXe century, the notion of resources also played an important role in the French geography school inspired by Paul Vidal de LA BLACHE.

The cumulative tradition of knowledge gradually faded in geography at the same time as the discipline became more illustrative. Despite everything, it persisted until the mid-twentieth century^{e} .

After the Second World War, the concept of the resource retained a prominent place in the discipline. However, the environmental impact of economic growth gave a new shape to the analysis of natural resources. The economic sciences are profoundly renewing the terms of the debate. The relationship between economic thought and the problems of resources underwent profound changes between the beginning of the XIXe century and the end of the XXe century.

Today, we feel that the concept of resources underpins interdisciplinary analysis and conditions the content of any attempt at modelling.

1-2-4-1-3-1- economic approaches

For economists, to give a definition to the concept of resource, we must necessarily refer to the analysis of value. For them, when something is competitive, attractive or coveted, it automatically becomes a resource. (For example, water, forests, even furniture).

The economist Le Robert defines resources as "the material means of existence". For him, if these means were in perfectly unlimited quantity, equally accessible and free of charge, everything that exists would constitute a resource; there would be no problem either in defining the concept, or a fortiori, in managing resources. However, this is not the case. To avoid any confusion, it is therefore necessary to clarify the concept.

It is very difficult to find definitions of the concept of resource in the writings of economists. They use related concepts depending on the economic and social context, which can implicitly be replaced by the concept of resources. So, to define a resource, we need to question the relevance of its evolution.

For the physiocrats of the XIXe century, and in particular in Francois QUESNAY's Tableau Economique (1759), the only source of wealth is land, which alone produces value, and therefore a net income supplement.

Francois Quesney's definition is inadequate, since the author only considers land as a resource, excluding forests, water, energy and many others.

For Adam Smith (1776), the wealth of nations is constituted by a flow that could be likened to the national income produced during a period or, more precisely, by consumer goods.

In Adam Smith's postulate, the notion of time and work appears. For him, a resource is produced by man's work, and this is done within a specific period of time, according to the needs that arise.

For Thomas Robert MALTHUS (1798), the means of subsistence are limited, so irrational use would lead to the risk of extinction. By 'means of subsistence', we mean the resources that nature provides for mankind (forests, land, water, energy, etc.).

David Ricardo (1819) developed the idea that the scarcity of resources was the cause of the end of economic growth. For him, a country's economic growth depended on the resources at its disposal. When these become scarce, economic instability is born.

For Thomas Robert Malthus, the limits to growth lie in the cost of using resources that a society can bear; for the Ricardians, there are no absolute limits to the scarcity of resources but only relative limits linked to the increasing cost of extracting and making resources available.

According to Jean Pierre BLOT et al (2004), in their book 'Ressources', the notion of resource is a social construction and a pre-existing datum, which becomes an interface between elements of the social order and elements of the natural order. According to this approach, every resource is the product of a relationship: "*there is no resource in itself*". For these authors, a resource is not primarily the product of an economic relationship, which would attribute meaning to it only through its market value.

Jean Pierre BLOT (2004) defines the notion of resource as "a dynamic relationship with space, resulting from the interaction between a group of actors and physical space" (P4).

For the economist, the concept of a resource can only be understood in terms of the functional relationship between the elements that produce it and the society that uses it. This relationship depends closely on the technology, the state of knowledge, and the economic and social context [...] so the ecosystem is not a resource, it is merely its condition of existence; an element of the ecosystem only becomes a resource when it has a value and when we take advantage of it (Jean Pierre BOURD et al, 1993, p 279).

For a long time, economists regarded the earth's natural resources as inexhaustible, and as a given once and for all. The main consequence of this theoretical assumption was a profound lack of attention to the impact on the environment, and in particular on the rates of reproduction.

1-2-4-1-3-2- physical approach

According to physicists, the concept of resource does not concern the materials necessary for human life. For them, the definition of a resource changes every time humanity transforms its technology, every time it introduces new physical principles, a new mastery of natural forces. For example, in ancient Mesopotamia, the resource 'petroleum' or 'rock oil' was considered a curse for any farmer who saw it emerging on his land. In their premise, they raise the question: is oil as useless today as it was yesterday? In support of their argument, they say that "the increasing usefulness of oil today can be explained by the fact that man has been able to penetrate the secrets of nature". In their view, by interacting with nature, man has mastered certain types of product (chromium, zinc, gold) that it contained within itself. He developed them, passed them on to his brothers and used them to improve nature.

Mining, forestry, shipping and agriculture still depend far too much on the vagaries of nature. Some nations are rich in these resources, and cruelly lack others, and vice versa.

Thanks to the mastery of nuclear energy (fission and soon fusion), humanity is already capable of eliminating the alleged global shortage of fresh water and irrigating any part of the world. In the field of minerals (metals, etc.), a large number of laboratories around the world are in the process of developing controlled atomic transmutation techniques to obtain one material from another. So, when all these projects have reached maturity, humanity will be free from the vagaries of nature: wherever it decides to stop, it will be able, in principle and with the help of living beings, to produce everything it needs from the materials directly within its reach. So, as long as mankind exists, 'resources' are inexhaustible .[25]

Their reflections revolve around man's capacity to create. This reflection seems relevant to us in that they emphasise human creativity. For them, nature's only resource is human creativity, and that is the only one we have to worry about. In their analysis, we find the notion of the human resource, which they describe here as the primary resource.

1-2-4-1-3-3- Types of resources

Whatever approach is used, there is a diversity of resources, including material resources, immaterial resources, heritage resources and moral resources.

Material resources include natural resources, of which forestry resources are one, and human resources. For the sake of accuracy, our analysis will focus on natural resources, with a briefing on human resources.

1-2-4-1-3-3- 1-Natural resources

Natural resources are the elements of nature available to a community for living and developing economic activities. In other words, natural resources are the various mineral or biological resources necessary for human life and economic activity. In a way, a natural resource is a substance, an organism or an object present in nature and which is in most cases the object of increased use to satisfy the needs (energy, food) of humans, animals or plants . [26]There are two types of natural resource: renewable and non-renewable.

Renewable resources are resources that can be replaced relatively quickly over a human lifetime. These resources are often both energy sources and elements essential to human life, such as water, sun, wind, the Earth's internal heat flow (geothermal), plants (biomass) and wild animals. However, over-consumption or degradation can lead to their disappearance or hinder their use (animal and plant species, primary forests, excessive pumping and/or pollution of lake and ocean waters, etc.).

Non-renewable resources include mineral raw materials (bauxite, silver, sand, potash, phosphate and diamonds, etc.) and fossil fuels (oil, bitumen, gas, coal,

---

23www.larecherchedubonheur.com/article-10507175.html

[26]*(http://www.fr.wikipedia.org/wiki/Ressource_naturelle)*

etc.), which come from deposits formed over the course of the Earth's geological history and represent an inherently finite stock.

Natural resources also include those that have not yet been exploited, commonly referred to as "reserves". Reserves are generally used to describe non-renewable resources such as ores and fuels. Given the technical possibilities of extraction and the financial conditions, they can be exploited at some point in the future. There are two types of reserves: certain reserves, where the deposits are clearly identified and delimited, and probable reserves.

According to the "replenishable/irreplenishable" criterion, natural resources can be classified into three categories, including economic value, vitality and renewability [27] .

Economic value

According to classical economists, a raw material can be considered a resource once it has acquired an economic or market value.

Vital resources take into account certain more or less vital characteristics. These include oxygen from photosynthesis, drinking water and everything we eat, as well as a large proportion of energy sources (fossil fuels and biomass), medicines, textiles and paper.

Renewability

This typology is based on the requirements of sustainable development. It classifies natural resources according to whether or not they are renewable. The aim is to ensure that resources are used sustainably, in line with the environmental, social and economic criteria of sustainable development.

1-2-4-1-3-3-2- Human resources

Human resources, or HR, originated with economists, and the concept dates back to the Industrial Revolution in England. They sometimes use the expression "human capital" to describe it.

Since Adam Smith, most economists have recognised that the skills of a country's workforce are one of its most important competitive assets.

Human Resources is a company department headed by a Human Resources Director (or sometimes by the Managing Director or the Finance and Administration Director in smaller companies), who is responsible for the administrative management of the workforce .[28]

According to the OECD definition[29] , human capital covers all the knowledge, qualifications, skills and individual characteristics that facilitate the creation of

---

[27] For this classification, we draw on the work of Georges ROTILLON, entitled "Economie des ressources naturelles", 2005, 124 p.

[28](http://www.jobintree.com/dictionnaire/definition-ressources-humaines-123.html)

[29] Organisation for Economic Cooperation and Development.

personal, social and economic well-being. It is an intangible asset that can enhance or sustain productivity, innovation and employability. Joseph STIGLITZ puts it this way: *"Human capital is the accumulated skills and experience that make employees more productive" (2007, p190).*
Samuelson NORDHAUS, (2000), adds that it is "the stock of technical knowledge and skills that characterises a nation's labour force and is the result of investment in education and training".

1-2-4- Analysis model

In this study, we adopt an analysis model from which a questionnaire will be drawn up. We believe that this will facilitate the fieldwork. By analysis model we mean a structured and coherent set of concepts, with their dimensions and indicators, which are linked together.

Raymond QUIVY et al (1995) define the analysis model as being the natural extension of the problem by articulating in an operational form the points of reference and the avenues which will be retained to preside over the work of improvement and observation. It is made up of concepts and hypotheses that are strictly articulated to form a coherent framework.

In their view, building a model of analysis means developing a coherent system of concepts and operational hypotheses.

We believe that conceptualisation is one of the main dimensions in the construction of the analysis model, because without it, it is impossible to imagine work that is not lost in vagueness, imprecision and arbitrariness. According to Raymond QUIVY et al, conceptualisation is an abstract construction designed to give an account of reality. It retains only the essentials of the reality concerned from the researcher's point of view; it is a selective construction.

This work therefore requires us to design the dimensions that make it up and then to specify the indicators by means of which the dimensions can be measured.

These authors distinguish between two types of operating concept. On the one hand, there are isolated concepts which are constructed empirically on the basis of direct observation or gathered information. On the other hand, there are the systematic concepts, constructed by abstract reasoning and characterised in principle by a greater degree of rupture with prejudice and the illusion of transparency.

As emphasised above, the purpose of this model of analysis that we are adopting is, on the one hand, to determine and characterise the performance of natural resource management regions and their impact on people's lives; and, on the other hand, to compare the regions with the socio-economic reality of the people who derive their livelihoods from their resources.

To do this, we have chosen a participatory approach, as this takes into account the empowerment of local populations and their participation in decision-making bodies.

Our hypothesis establishes relationships between three concepts: The concept of actors, governance and resources.

DESIGNS	DIMENSIONS	INDICATORS
Governance	-power sharing	Community participation in decision-making ; -Taking account of people's needs
	Developing and sharing environmentally-friendly techniques and technologies	-Adapting tools to the needs of the population -Mobilising management tools
	-Promoting rural finance	Encouragement of off-farm income-generating activities - Reducing pressure on forest resources -Empowering local communities
Players	-Institutions -Local players -Status	Support for village communities from imposed areas without taking account of the socio-economic environment (outside the community rules) -Each participant must be familiar with the content of the rules drawn up; -Village Cooperative - Involving local people in the management of their resource ; - Draw up their own management rules - Individuals organise themselves to solve common resource problems, but appropriators have always established rules that limit the actions they are allowed to take; - Transferring power to communities; - This transfer of power must be accompanied by capital; - Commitment by the State through measures to protect the massif ; - If the measures selected are too repressive, this will
	-Involvement of players	can lead to conflict between the different parties (administrators and communities); - Efficient involvement of all decision-makers at all levels; - If everyone, or almost everyone, follows the rules, units of resources will be allocated in a predictable and efficient way; - Conflict levels will be reduced;

		- The resource system itself maintained over time;
Resources	-Free access	- There is no limit to the number of appropriations; - Independent ownership ; - Endemic dissipation of resources; - No one has the slightest incentive to leave units to others; - Physical violence occurs; - ways of resolving violence ;
	-Acces fermë or limit	A well-defined group is the owner; -Joint operation -Monitoring ; -Owners face incentives that depend on the rules governing resources

CHAPTER II

THE ZIAMA FOREST MASSIUM: an an ecosystem with diverse and complex issues

The sustainable management of a resource requires the establishment of principles or laws that serve as a basis for guaranteeing and preserving the ideal set for this management. These principles should be implemented and monitored to assess the impact of actions on local populations and the ecosystem in question. According to Barre KOIVOGUI 2014, "*you cannot manage a forest with all its resources without first having a management plan*" 8^2, which means that the management plan is a document in which the broad outlines of these principles or laws are mentioned for better organisation of actions in order to rationally manage the resources that a forest contains. He adds: "*The management plan is still not well known by Guinean foresters* [30] [31]".

As David BOLLIER (2012, p12) puts it: *"the greatest value of common pool resources is their ability to help us assert another axis of value [....]; so I see common pool resources as highly generative. They create all sorts of values - material resources, social connection, a sense of identity and belonging"*.

With this in mind, the Ziama forest has had a management plan since 1991, in which strategies have been developed to identify resources, plant forms and the concerns of local people, with a view to participatory and concerted management through a planned and executed inventory.

2-1-From village land to classification: a succession of different changes

The management of Ziama's forest resources, like any other classified forest, should be based on knowledge of the relationship between the forest and the neighbouring villages, i.e. knowledge of the history of the region and the different relationships that people had before it was classified. According to our analyses based on the writings of James HEATHER (1994) and our field surveys, most of our classified forests today were once places of habitation. The existence of the ruins of old village sites in certain parts of the forest testifies to the presence of earlier peasant occupation of these areas in relation to the classification of the forest. One example of ancient occupation of these forests is the assembled stones that served as foundations or places of worship for certain villages.

---

[30] Taken from the interview given to Mr Barres KOIVOGUI, Ziama branch manager during our field survey on 21 August 2014.

[31] A forester may be a person who exploits forests for economic purposes, in particular for timber production. In this case, he is called a silviculturist. For the purposes of our research, a forester is an employee responsible for maintaining and looking after forests, particularly public forests.

Zolea KALIVOGUI [32] [33], referring to the early occupation of this forest by farmers, said: "*Our parents cleared this forest and at the same time helped to regenerate it by planting cheese trees around the village*".

When the colonists arrived, they asked for the forest to be classified, with the agreement of certain villages. However, others refused. So a system was adopted for them, the "forcing", adds Zolea Kalivogui, "*there were villages in the middle of the forest, and with the forcing system, these villages left to regroup around Guebe. 31*". Depopulation caused by tribal wars, the colonial conquest and disease were all factors that would explain the abandonment of certain villages in Ziama.

The Ziama massif covers the Waima , [34]made up of seven (7) Tomas villages including Dandano, Paran, Goboëla, Koima, Irie, Boo and Baimani.

To the west of Waima was Ziama, once a kingdom whose capital was Boussedou.

Ziama comes etymologically from "Ziema" of Toma origin, which in turn derives from two words "Zie" meaning (a marigot) located in Kouankan. This marigot has long been a source of vengeance for the local people, and served as a source of power for the kingdom of Boussedou. And the word 'ma' means 'beside'. So Ziema means "beside the marigot". Ziema is the territory crossed by the Zie marigot. In this marigot lived the protective genies of the whole kingdom. During the tribal wars, it was annexed by Samory Toure's Wassoulou kingdom. At the time of this annexation, the population fled. Some of the inhabitants fled to Liberia and the mountains. Finally, Ziema refers to a people that stretched from the Gurnee (Balloma, Sangolome, Kanssanka, Ngnalessou, Dopame) to Liberia, precisely the village of Zegne. In Toma country, some people write it as "Zyema" or "Ziema", but with the franchise transcription, it is now written as "Ziama".

The disruption caused to the forest by all the above-mentioned events is said to have been the main reason why the inhabitants moved to other locations. Today, the inhabitants of this forest remember the former occupations of their ancestors. They are demanding the return of the land they once held, which they consider to be a reserve.

---

[32] Zolea Kalivogui, one of the founder's grandsons born in 1920, testifies to having seen the forest before and during the classification.

[33] Guebe was the founder of the village of Irie, which was created in 1770. The culture of hospitality that he instilled in his children enabled them to welcome the villages that were evacuated from the forest when it was classified by the colonists in 1942.

[34] Waima was a kingdom with Koima as its capital, and its king was called Dowilnya BILIVOGUI, located to the south of Kouankan. Today it refers to the southern Toma whose language often uses the syllable "gui".

2-2-Ziama, a theatre area between 1868 and 1909

From 1868 to 1909, the Ziama region was the scene of violent tribal wars and colonial conquests between the local population and the colonists. This region, once densely populated and prosperous through trade, fell into decline.

Before the colonial conquest, the region was a major trading centre between the populations of the north (the Malinkes), the south (Toma, Guerze) and the coast. As for the Malinke traders, they were always present on the markets of the Ziama region. They eventually settled permanently in the region. This is how a Malinke kingdom called Bouzie came to be established on the right bank of the Diani river, with Kouankan as its capital.

Even before the colonial conquest, the Toma country was decimated by tribal warfare, especially the Waima Tomas region, which was in conflict with the Malinkes (James HEATHER, 1994).

These wars were fatal for the Tomas, as James HEATHER points out (1994, p13): "*A large number of dependent villages that remained from the XIXeme century have completely disappeared. Boo has lost 21 satellite villages and Koima at least 4; Kothia, an important fortified village from the XIXeme century, which itself had dependent villages, no longer exists. It is located in the strictly protected sanctuary of the modern Ziama reserve*.

2-3-Classification of the Ziama forest

After the total devastation of the Waima Toma country by the colonists, they occupied the Ziama region.

Civilising conservation missions visited the forest. The first visitor arrived in 1909, and described the secondary nature of a forest

pushing back a ruined, decimated Tomas region onto former farmland.

Thirty years of human absence have allowed the forest to regenerate, and he describes the green vegetation from Soundedou to Fassanconi, noting a certain scarcity of large trees and creepers.

The forest was then visited in 1942 by the botanist and forest administrator Adam.

As mentioned above, these wars of succession and migrations to the south decimated the Toma population and destroyed many of their villages. Presumably, the Ziama secondary forest lasted for forty (40) to sixty (60) years overall, the date of the Malinke intrusion.

In 1932, the inhabitants of the Ziama region had their territory reclassified for the first time by the French colonial administration. This was the first classification of this reserve.

During the demarcation of this classified forest of 119019 ha, the colonial administration solicited the participation of the various riverside villages that

already existed. Together with them, they drew up the boundaries according to their ecological importance, which at the time concerned mountain ridges and watercourses. This is how the boundaries were drawn, taking into account ecologically sensitive areas. It so happened that within this forest there were enclaves and riverside villages, of which there are twenty-eight.
The current problem is that, from a human point of view, it is difficult to move these people from the forest without prior support measures, as they have every right to live there. Aware of this reality, the colonists found another strategy that would enable them to keep the forest classified and allow the population to live off the resources of their environment. The method adopted for this was that of "enclaves". It consisted in knowing the dynamics of the population concerned, its growth over long years and fixing on a given area that should serve it.
We know that there were 500 inhabitants in Boo in 2014, for example, and there should be X in 2020 and Y in 2050. Finally, the question arises as to whether all these people can live within this enclave at this rate of growth?
It seems to us that no forecasts were made during the colonial era. If there was, its impact was not visible on the ground, because the populations of these enclaves have increased considerably and far exceed the agrarian capacity of the village terroirs that once existed. This situation has long prompted these populations to claim ownership of the forest for agricultural purposes. Barre KOIVOIGUI came back to this problem in the following terms during our field survey: "*The fundamental problem that I have today for the management of the classified forest of Ziama, is that of enclave, I have to find a solution one way or another. I don't have the solution on my own. However, I'm considering and proposing solutions to the government through my superiors, and while we're waiting for the problem to be solved, which will be difficult and even impossible, we need to look at ways of enabling these guys to live happily at home without overflowing the cages of their enclave.*
Massa Guilavogui goes back over the reasons for the forest's classification in the following terms: "for *the colonists, the forest had to be classified to prevent the massive descent of the Malinkes to the south, who were cultivating the savannah. They feared the destruction of a wild forest that was supposed to be classified".*
Behind this idea was an economic goal for the colonists. To conserve forest species in order to support the burgeoning industries in Europe. It should be noted that economic reasons were the main reason for the classification of forests in the various colonies.
The act of classification was official in 1942, in April to be precise, under the authorisation of the Governor of the AOF. In addition, it was extended for the

production of quinine, because the region was far from the coast for logging purposes. Tea plantations were created in the forest.
According to the ancients, the colonist was carried by the blacks in a hammock when the forest was demarcated.
At Seredou, for example, the boundary was drawn along wele . [35]The settler's porters were tired. They didn't want to carry the master through the forest any more; they would draw the forest line not far from the village in some cases, such as Zoboroma.
In 1943, the Tomas realised that the classification of the forest served only the economic interests of the colonists, and not to prevent the mass descent of the Malinkes. For them, this was a loss, as their socio-economic life depended on the upkeep of the fallow land, which was regularly renewed for agricultural purposes. Today, Ziama is becoming increasingly important in ecological terms, no longer a local or national issue but a global one, as it is the focus of major international projects. It was classified as a world reserve in 1981 by UNESCO. The very first conservation project was financed by the World Bank in 1988.

2-4-Between the laws governing classified forests and management practices, a recurrence of various conflicts

Limits on access to resources and indigenous regulations cannot be truly understood if we neglect the means used by those who control them to take advantage of them. All modes of prohibition, control and limitation are placed under the vigilance of customary authorities. A frontal study of these methods can only yield limited results. This is not to say that the surveys do not want to reveal the truth, but many of the prohibitions are not explicitly stated. They are obvious to the entire village community because they are part of the strategies of the households, the lineages and the village. It is therefore by studying all these strategies that we can gradually shed light on the different ways in which resources are managed. Indeed, resource management is an exogenous concept in African society. Pascal REY (2007, p194) points out that '*the notion of resource management is specific to our Western societies and is not necessarily understood in the same way by African societies*'. Indigenous management methods are not based on a body of rigorously linked principles, but rely on ideology and the sacred, and consequently on '*an element of arbitrariness, prohibitions and concerns that cannot be rationally explained*'.

2-4-1-Customary practice: the primary method of managing the Ziama forest

Throughout history, societies have invented and developed, mainly on a local

[35] Wele is a river just inside Seredou when you leave Macenta, it crosses the sub-prefecture in its western part.

scale, ways of managing natural resources collectively. These methods were one of the foundations of their survival and wealth (Olivier PETIJEAN,2010). In some cases, it was a question of managing the relative scarcity of these resources and preventing the conflicts that this could engender. The choice of a form of pooled management was based on the simple observation that it would enable a greater number of beneficiaries to get more out of the resources available, while preserving them for future generations and ensuring the conditions for the perpetuation and renewal of their societies.

Based on our fieldwork, it seems that most of the villages bordering Ziama existed before the forest was classified. However, the management method that prevailed before the arrival of the colonists in the region may have been customary. Through the testimony of the elders, we can see that people lived in this forest and maintained links with it in one way or another.

At that time, forest resources were managed on a customary or traditional basis, governed by community rules. Sehve KOIVOGUI points out that "*according to custom, there were prohibitions established in all the villages, stipulating that: no one was to cultivate until they reached the rivers, the mountain tops or near the villages; children were not to wander in the forest. As for hunting, it was closed from June and September until December. This was their way of preserving their heritage. Anyone who exceeded these regulations had to pay a fine in kind. For example, a can of red oil, a billy goat, ten kilograms of rice and a flask of white wine. If he refused, the person concerned could be isolated by the community, no one would visit him and if he had products for sale no one would buy them. If he was married, his in-laws would take his wife away from him. All this was a way of bringing him to his senses*". In the eyes of many researchers, this is seen as the very first way of managing natural resources. Pascal REY (2007) points out that traditional power in the management of natural resources is a significant factor. For him, the terms 'custom' and 'tradition' refer to a heritage of the past whose origin is not clearly circumscribed in space and time, and which has been passed down from generation to generation. As far as we are concerned, traditional or customary power is defined by the reiteration of the founding act. It is based on facts and beliefs inherited from ancestors.

According to Pascal REY (2007, p30), "*traditional powers are based on the sacred land, the dead as glorious illustrations, symbolism derived from founding figures and the time of origins, and the political liturgy governing the practice of power*".

On this subject, Koi Guilavogui emphasised that "in *each village there were cult forests for men and women, forests where the genies lived and which were reserved for their adoration. These places were off-limits to all activity, and were a way for them to conserve the forest. In fact, they made it possible to have the natural forests that have remained untouched until now in some villages. As the villages grew, the ones that were there were moved to other places".* These traditional methods of conservation were nonetheless gaining ground in these villages in order to maintain the forest's resources. It should be noted that, at the time, the dense forests formed the belt around the villages bordering Ziama, as each family worshipped the cheese trees because they carried protective and rain-giving genes. The FAO describes the same system *"in the Agdal forests*[36] *of Morocco, the local communities establish the rules governing the authorised periods, as well as the volumes and species that can be harvested; offenders are required to pay a heavy fine to the local community"* (FAO, 2012, p13). This system provides an overall conceptual framework integrating the ecosystems and natural resources of a territory, knowledge and uses, rules and institutions, and perceptions and beliefs. Even in Namibia, forests and wildlife were traditionally protected as part of rituals and sacred actions.

Most of the elders we interviewed during our research testified to the effectiveness of this management method. According to them, it enabled better conservation of the forest's resources. In this regard, Marcel Sow (2003) notes that the traditional management system has, since ancient times, enabled the regeneration of forest resources; management based on religious instructions, formulated by the village elders. It was respected to the letter by all the villagers.

2-4-2-State management practices

During the colonial period, the protected areas were managed by the colonist who represented the metropolitan state in the various colonies. In the specific case of Ziama, this method can be divided into two periods.

2-4-2-1-The period from 1942 to 1990

Our forests, whose classification began in 1932, were managed directly by the colonial administration. To justify their presence, they built housing estates inside the forests. In the specific case of Ziama, it seems to us that these cites served as learning centres for the notions of nature conservation and protection. In some cases, they were used as living quarters for the conservations, to better watch over the forest and its resources. Some even exist today, including these.

[36] The term 'Agdale' is widely used by pastoral societies in North Africa to designate an area, a resource and the rules laid down for its management, but here we borrow from the FAO's examples for Morocco.

Historic town of Ziama
Source: Candidate survey, September 2014

In Africa, forest management, as an organised and planned process of exploitation, dates back to the first half of the 19th century, mainly in West African countries (Maya LE ROY, 2013). The principles of logging were largely supported by a colonial ideological corpus. In order to protect Ziama's natural resources from exploitation by indigenous peoples, the colonial authorities took strict protective measures at an early stage. However, the measures taken at central level and the practices of the peasants in the area did not coincide. But these forestry policies - the law, the forestry code - were imposed, and disrupted the socio-economic organisation of the local populations. These conservation practices were not accompanied by respect for customary rights. The local people continued to exploit forest resources according to their traditions, which more often than not were at odds with the rules of public administration[37] . These policies include [...] the construction of a technical forestry corps (creation of colonial forestry services), the development of forest science teaching and research services; the delimitation of forest domains by ensuring access to resources and their control by the state; the development and use of forestry technologies and harvesting practices (Maya LE ROY, 2013).

A sawmill was built close to the forest to produce timber, a large quinine plantation was opened and a factory was built to produce tea. All these facilities were used to exploit the resources of the forest, which was a source of income for the colonists.

2-4-2-2-The period from 1958 to 1990

After the departure of the colonists, i.e. following independence, most of the colonial property reverted to the State, which was the new master of the young State. The system for managing natural resources in general and land resources in particular remained virtually unchanged. Indeed, the laws were modelled on

[37] (www.unesco.org/mab/doc/savanna/doc/RapportFinal.pdf

those of the West throughout Africa. In the eyes of many observers, this was the colonial heritage. For some, it was a continuation of the colonial management system, as Dilys ROE, 2009, p 23, puts it. "*Ultimately, this evolution of land tenure became one of the main drivers of African independence movements seeking to regain their rights to land and resources. Resources such as wildlife were gradually placed under a central regulatory regime, and the rights of local populations to use resources were gradually alienated*".

In line with this logic, the authorities of these new States have largely copied and pasted Western policies in all areas. As far back as 1968, authors such as Garret HARDIN were already encouraging "state" control of the bulk of pooled natural resources in order to prevent their destruction. Martin YELKOUNI (2004), for example, justifies the conservation of these resources for their ecological role. In the specific case of Africa, they are part of a set of resources managed by the state.

After independence, land rights in many countries, such as Ghana, Mali, Côte d'Ivoire and Gurnee [...] instead of being decentralised, became even more centralised.

When the Ziama forest was classified in 1942, even after the colonists, it continued to retain the status of a classified state forest. To this end, Barre KOIVOGUI points out that, "*all that could be done was the fact that everyone knew that this forest was classified by the State and that no one should destroy it. So there was talk of planting Chinese bamboo to serve as a boundary, and these species grew to make a good belt. But no other practical management strategy has been implemented, with the exception of surveillance by state agents.*

From the 1970s onwards, public policies came up against difficulties in succeeding, because these resources were used by a large low-income population, or for their competitive use, so the forest came under various pressures. In addition, the state was no longer able to control negative externalities because of budgetary constraints. As a result, the forest was over-exploited, leading to its degradation. The local populations no longer felt concerned by the concessions because they were not involved in drawing up the regulations. As Sehve KOIVOGUI points out: "*People were often caught in the forest taking species. In such cases, the forest rangers would take money from them and they would continue on their way. In the end, the forest became like a panther's skin, where everyone paid something to get a plot of land in which they could plant all sorts of crops*. Since it was declassified and negotiated to be individually appropriated, and then allocated to agriculture, anyone who could afford it could get it.

The law, which stated that the land belonged to the person who developed it, gave the population free access to the forest and its resources by negotiating with the forest guards for its development. Many people settled in the classified forest to open vast coffee, cocoa and cola plantations, particularly the Malinkes, most of whom came from Beyla. At first, they were contract workers, but eventually they settled down. At Avillissou, for example, Moussa TRAORE [38]had this to say: "*We came from Beyla through the channel of our brother Mamoudouba SOUMAORO at the time of Sekou TOURE. At that time, we worked in this forest without any conditions and we grew cocoa and coffee. When Lansana CONTE came to power, we were chased out of the forest in 1990*. Koi Guilavogui also underlines this fact in these terms: "*There was also one at Sibata II, where Mr Grovogui had asked the government of the day for part of the classified forest to be used as a plantation. When he moved in, some members of his family left Sibata I to settle there. This was followed by Ballassou in 1979, with the occupation of the current sanctuary.*

This weakness on the part of the State has allowed many people to attack the forest once again. It has led to the degradation of a large part of the forest, around 3,000 ha, across the neighbouring village lands.

We believe that state management has not really borne fruit because it was all about surveillance and not the application of measures. Negotiating agricultural plots with the forest guards of the time within the perimeter of the classified forest was a daily practice. It was in the light of all these attacks that the forestry administration realised that it needed to change its approach. Otherwise, in the next few years, we would no longer be talking about the classified forest of Ziama. A farmer could only enter a classified forest with permission from the forestry services. At the time of our investigation, official documents showed that third parties had been given permission to settle in the classified forest.

2-4-2-3-Conflicts as indicators of the evolution of a crisis in the management of common resources in Ziama

The Ziama forest was once the embodiment of the cultural identity of the riverside communities. But since 1942, these cultural links have been increasingly weakened by the classification. This change has led to a constant conflict of interests between the various players involved in the life of the forest. Since 1992, everyone has been trying to make the most of the forest's wealth of flora and fauna.

On the one hand, the communities are claiming ownership of their property, which for them remains an inalienable right, and the exploitation of its resources

[38] Moussa Traore is one of the wise men of Avillissou from the Mania community whom we met during the survey on 23 August 2014.

is essential to their subsistence. On the other hand, the forestry administration is making concessions that are contrary to the expectations of these communities. For the administrators, the forest must be protected against any aggression that could endanger the life of animal and plant species. On the other hand, the communities believe that the forest is a "gift from God" from which they should draw their sustenance. They should therefore be left to manage it in their own way.

The claim to ancestral lands in the Ziama region is nothing new. It dates back to the earliest days of the legal deed to which the cultivable land of the Tomas was subject.

For a long time, there were two diametrically opposed visions: on the one hand, a virgin forest in wild Africa that had to be conserved to ensure its continued existence, and this conservation required classification. On the other hand (among the Tomas), this forest, considered wild and classified for its conservation, invaded farmland and fallow land. What's more, the depopulation and ruin of the region by the colonial heritage was a perpetual symbol of desolation and loss of prestige; abandoned dwellings and fields overgrown with wild vegetation reminded the villagers of their defeat.

The villages surveyed after the colonial conquest maintained their claim to the land. Their descendants continue to put forward these demands concerning their ancestral land.

These claims are based above all on the status of the founder of the family. For the Tomas, even if their ancestors are dead, the villages of the past have been abandoned and fallen into ruin, but the land still exists. "*This land is our heritage, that's what we're asking for",* says Koikoi Guilavogui.

In response to these conflicts, the reserve's boundaries had to be modified and it was reclassified in 1943, giving local people more land to cultivate close to their villages. For the colonial administration, local interests were satisfactorily taken into account; it saw its interests in terms of immediate agricultural needs rather than long-term territorial claims.

Following independence, the forest reverted to the State. Under the first republic, there was widespread encroachment as a result of the weakness of the forestry services. This weakness between 1970 and 1980 enabled Malinké immigrants to obtain land that had been legally or semi-legally ceded by forestry officials. However, these methods of acquisition ran counter to the Tomas' concepts and relations concerning land allocation rights. Siba Kalivogui had this to say: "*They took our land and sold it to people we didn't know. In our country, only the first people to arrive have the authority to allocate short or long-term rights of use* to *third parties*". This marked the beginning of a new tension

between the Tomas and the Malinkes, and also between the Tomas and the conservatives.
Today, we can see the active commitment of all the local communities, under the motto "*our ancestors settled in this region, cultivated it and worked for its productivity*".
The Tomas' claim is far from short-term, insofar as they are talking about territories once occupied by their ancestors.
In the course of our survey, we identified three types of conflict: conflict between animals and the community, conflict between the community and the State, and inter-community conflict. These conflicts will be illustrated in the following tables

2-5-1-Conflict between animals and people

The classification of the forest has led to an increase in the number of animal species in general and elephants in particular. The animals can no longer find food within the 119019 hectares. They often go beyond the boundaries of the forest and enter village lands in search of food. The most common is elephants, which threaten the survival of the local population, who are forbidden to touch most of them in some places. These fully protected animals ravage fields and even harvests. As Koi Guilavogui explains: "*In 2012, I found that elephants had caused enormous damage to many fields in Boo, in response to a request from local people. I then submitted a report to the government through my superiors, but they never came to the aid of the victims.*
There is still a great deal of material damage in some villages, but the loss of human life is also very worrying. In Boo, for example, three victims of this damage were reported to have died chasing an elephant.
The upsurge in this conflict began when the elephant population increased in the forest, probably during the Liberian war in 1989. Sehve KOIVOIGUI commented: *"This conflict began with the first Liberian war in 1989, which drove elephants into the Ziama massif. Some of them were domesticated, like the elephants of Liberian President Samuel; as the park was ransacked, they came out to invade the forest. At that time, some elephants even went up to Irie during the day and at night. They would sometimes pass through the village".* In 1999, the elephant population was estimated at around 141 individuals (Sarah SANDERS, 1999). Today, the number of elephants has risen to 214 (CFZ). With this number, there is naturally a need for more food, which pushes these elephants beyond the limits of the forest. These elephants have propitious periods during which they leave the forest in search of food. Sory KOUROUMA

[39] explains: "*Elephants are present from September to January to ravage our fields and at the time of ripening. We used to be able to produce six tonnes of bananas every market day, but now we've all given up because of these elephants. We want Alpha Conde to chase these elephants away, because they're making us tired. Our small fields in the lowlands are being destroyed, and that's what's causing the hunger gap. These elephants don't stay in the classified forest, they roam all over the place and we have no right to touch them. At least if we could keep them in the forest, we would be at peace*.

According to the testimonies gathered in Irie, elephants are a nuisance from the sprouting stage of the rice fields to the harvest. They even destroy the granaries in the fields. And yet these granaries are used not only to store food, but also seeds for the following season. The harvest is kept for threshing and transported bit by bit, given the distance between the fields and the villages. These elephants are frequent, but especially during the premature period. During the dry season, they come to investigate, but they have never killed any, although these elephants have claimed two victims in Irie, one dead and another injured. The conflict between the elephants and the people living near the forest is of great concern to the local communities because of the force of the law that protects them. Gbade BILIVOGUI points out that "*you can't kill them; when they catch you, you'll spend a lot of time in prison. The big problem is that if you cut down a tree, you're excused, but if you're an elephant, if you're not careful, you risk spending a good part of your life in prison*".

More often than not, the damage caused by these elephants is reported by local people, but no measures are taken to compensate them. During our field survey, a large number of local people expressed their dissatisfaction with these elephants, as they cause damage to their crops in the vicinity of the classified forest. In some villages, such as Irie and Avilisou, the conflict had reached such proportions that villagers had written to the authorities asking for help. For some, if the authorities can't help them compensate for the damage caused by these elephants, they must put them in the enclosures. The human-elephant conflict is far from over, as the two sides do not have the same perception of the problem. The forestry administration believes that it would be impossible to keep these elephants in an enclosure. As far as they are concerned, the elephants knock over everything in their path. Barre KOIVOGUI says: "*The state doesn't have the resources. Even in the most developed countries, I don't think that's possible, an elephant is a buldesert, certainly, they've been in electrified enclosures. You can't stop them from reproducing either*". The management of

[39] Sory Kourouma is the president of the Avillisou district with whom we spoke on 23 August 2014 at 2.40pm.

this conflict is extremely delicate. It even pits the reserve managers against the villagers. According to the villagers, the managers are entirely responsible for the destruction of their fields. As Koi Guilavogui puts it: "*We can't tell the people to shoot these elephants because they are under international protection. They shouldn't even think about killing them. On the other hand, we are not the ones who allow elephants to ravage their fields.* Nobody wants to take responsibility for the presence of elephants in the fields. This implies that the conflict is not recognised by the state. The villagers want responsibility to be established and accepted, even if there is no immediate damage. This lack of knowledge leads the farmers to defend themselves in their own way in order to secure their crops.

Table 4: Conflict between the community and predatory animals

Types of conflict	Degree of recurrence	Duration (year)	Number of fields destroyed	Loss of life	Start of the conflict	State of the conflict	Villages
Ekphants - Men	Tres ёкуё	25	50	3	1990	Unresolved	1пё
Ekphants - Men	Tres ёкуё	24	100	0	1991	Unresolved	Avilissou
Monkeys - Men	Tres ёкуё	24	0	0	1990	Unresolved	Zoboroma
Agoutis - Man	Tres ёкуё	25	0	0	1990	Unresolved	Ballassou
Agoutis - Man	Tres ёкуё	25	0	0	1990	Unresolved	Zoboroma

Source: Candidate survey, August 2014

2-5-2-Conflict of ownership between the Community and the State

The conflict that exists between the State and the communities is over the ownership of this forest. There is a surprising disparity between the local perception of history, which determines the priorities of the villagers, and those of the conservationists. According to James FAIRHEAD et al (1994), there is an antagonism towards this forest that goes back to its establishment and cannot be understood or explained outside the historical context. This conflict dates back many years. Some place it just after the forest was classified. In 1943, the colonial administration declared a part of the forest to satisfy the demands of the local population. For others, like today's conservationists, the problem dates back to the time when Sekou Touré said that the land belonged to those who worked it. This was an opportunity for the peasants to settle in the classified forest and make it their property. "*It all started under the first regime, when it was a question of paying tax in kind, but they said that the land belonged to the person who developed it, so people just went in and planted fields. It was after that, under Lansana CONTE in 1991, that people started to be evicted, which created difficulties for us, it was hard, because the plantations were there,"* says Koi GUILAVOGUI. At present, these farmers are not even allowed to maintain the plantations in the classified forest. For the authorities, allowing these people to maintain the plantations would be an opportunity to expand their property. Farmers must therefore be prevented from tending plantations in classified forests.

Under Lansana CONTE, the state took over management of the massif with strict and severe controls. Draconian measures were applied. These included chasing farmers out of the forest and cutting down their plantations. In most cases, people were evicted. In this way, the state became the absolute owner of the forest, granting only usage rights to the peasants. The oldest man in the village of Zoboroma had this to say: "*The forest problem is complicated. If you see that we no longer have control over the forest, it's because of the forestry officers who come with papers to demarcate the forest. If we are given responsibility for this forest, we are prepared to act in the interests of conservation. If we were given full responsibility for managing the forest, we could give you the assurance that we would protect it. But as long as supervision is left in the hands of the state, we can do nothing.*

For the forestry administration, the local population cannot under any circumstances be fully responsible for the forest. In their view, they do not have the necessary skills to successfully safeguard the species in the reserve. In their view, the farmers are already abusing the law by cutting down forest species and slaughtering protected animal species. So entrusting them with the management of the forest would be detrimental to the conservation of plant and animal species protected by international law.

Anything that compromises the content of the forestry code in force for the protection of the forest is automatically a conflict. This conflict began when the forest was classified. It is for this reason that the forestry administration has taken steps to apply the code and draw up sustainable management plans.

The farmers have a different analysis of the situation, as Kolouba DOPAVOGUI [40]declares: "*Our return here was not concealed; we informed the authorities at the time and they accepted us officially. Nobody can say they don't know. When we arrived, they welcomed us with open arms, now they should help us to stay here with our children*".

Farmers are still claiming ownership of this forest. In Zoboroma, for example, a group of young people rose up in 2011 to go and farm in the classified forest. The prefect of Macenta and the red berets invaded the village. The young

[40] Koulouba Dopavogui is the president of the Ballasou youth group, who was very passionate as he answered our questions.

people's motto, as Maurice DOPAVOGUI points out, was "*we don't have any land to grow crops, the forest is there, it doesn't work, it's not done anywhere else, even getting wood to build our houses is impossible. However, this forest belonged to our ancestors and they cultivated* it". The alleged recalcitrants spent two to three months in prison in Macenta. They paid one million Guinean francs each for their release. Since then, no one has touched the forest.

The forestry administration is unaware of the presence of certain enclave villages, such as Ballassou. According to the CFZ, this village was established without authorisation. Yet the village claims to have obtained official authorisation from the government of the day in 1979, and moreover, the current site is where their elders left to take refuge on Balla mountain because of the tribal wars at the time.

The most serious problem in Ziama is that even in areas where crops are grown outside the classified forest, their use by local people requires a forestry fee, which is paid to the CF. This amounts to a million a year per village. Yet these people have the right to grow crops in the part of the forest known as the "protected forest".

Table illustrating community and State conflicts

Protagonists	Degree of recurrence	Start year	Villages
CF -population	Very high	1990	Avillisou
CF - Population	Very high	1991 2011	Zoboroma
Forestry cantonment _population	Very high	2013	Irie
CF - population	Very high	1990	Ballasou

Source: I'enquete du candidat, August 2014

2-5-3-Inter-community conflict

Today, the Ziama area is a perpetual conflict zone, in that not a day goes by without a conflict breaking out between at least two individuals from different villages in the middle of the bush. In respect of the use rights granted to the local people, the State had granted lowlands in the classified forest. These low-lying areas are often the subject of conflict between village communities bordering the forest because they are simply insufficient.

Farmers use the lowlands mainly for subsistence. These are natural enclaves whose exploitation, if carried out properly, should not contribute to the devastation of the forest. In this activity, which the FC considers to be a right of use, people fight each other over these lowlands, "*I was here before, and I'm here today, and so on*". The forestry centre, which is a government technical service for forest management, is sometimes involved in managing this kind of conflict. As in the case of Avilisou and Boo, two contiguous village areas,

Malaweta and Chefkolydou. In Malaweta, the situation is still serious, as the conflicts are internal between the local populations.
According to the testimonies gathered in Avillisou, it seems that the Avilisou-Boo conflict started in January 2014 and became openly apparent in July 2014. It was resolved the same month (see tab1eau in l'annexe). Amara Traore said: "*The land in dispute between our two villages is Boo's and not Avilisou's, which is why we asked the CFZ to resolve the dispute.* The CFZ does not have a mandate to resolve disputes, but it is often invited as the institution responsible for managing these areas. It brings the protagonists together and fosters their mutual acceptance of each other, and it is beyond them that the problem can be brought to the level of the Macenta judges. For the moment, they have not yet reached this level. In this area, the degree of recurrence of conflicts remains very high and varies from one village to another, with the exception of Baloma and Ballasou.

Breakdown of inter-community conflicts by duration in months

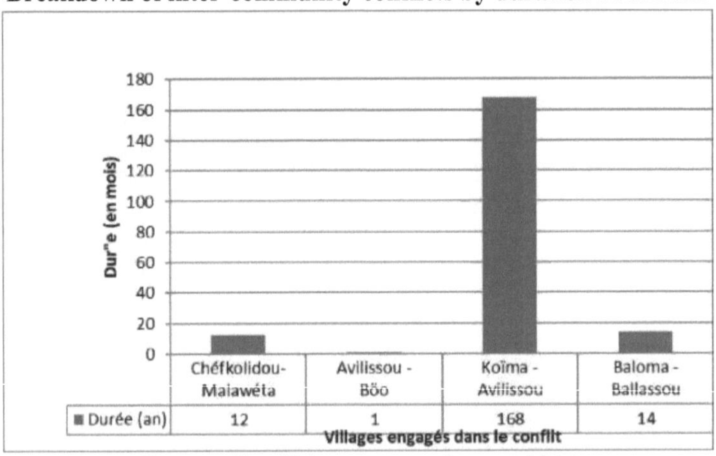

Source: Candidate survey, August 2014

2-5-3-1- The allocation of lowlands and their management in classified forests

Photo of a shallow area in the Ziama classified forest, taken by Marcel KOUROUMA, August 2014

The lowlands are selected according to the villages in need. A management committee is set up in the village. Its role is to allocate the lowlands to poor people on the basis of an agreement signed by the local authorities, who are responsible for reporting to the committee and the CR. This agreement sets out the technical guidelines. No lowlands can be obtained without prior verification, and lowlands are not individual property. They may be reallocated to other people the following year. They cannot be bequeathed from father to son. A symbolic quota marking state ownership is paid. This amount is fixed at 30,000fg. The lowlands concerned are geo-referenced and mapped. No extension of the area to be cultivated is authorised on the slopes. 1/3 of the amount is paid to the village committee for operating costs and 2/3 enables the forestry centre to monitor the progress of its work. Gbade KOIVOGUI commented: *It's not a profit-making activity, but rather a participatory management activity. The exploitation of these shallows does not take into account the origin of the people. Sometimes people don't pay their quota, and many of them overflow the cages and attack the hillsides. The measures mean that the interested party is dispossessed of the shallows, and this is what causes the conflicts. The conflict is often between the dispossessor and the taker. Often those who don't pay their quota are warned, and if they don't pay, they lose their rights to use the lowland to other people next season*. Failure to comply with the terms of the agreement would give the CFZ the right to take the land away from them. The lowlands are granted per family, and each family is entitled to two hectares. Often, this is after a check by the management committee. The management committee is chaired by the president of the management committee, who is followed by all the farmers. According to the UNCRPD (2012, P26-27), '*management committees must be made up of representatives of all social groups so that no group feels undermined and becomes the enemy of local forest governance*'. In this case, some citizens will be frustrated by the management system put in place because it will only benefit a minority to the detriment of the majority. This can be seen in Ziama, where the management committees lease the lowlands to their brothers for between 200,000gnf and 500,000gnf/hectare, while the CFZ officially leases them for 30,000gnf. This pushes some citizens to hide to cultivate in the middle of a classified forest. This often leads to conflict between the managers and the local people. As was the case in Avilissou and Sibata II.

The CFZ is generally faced with huge difficulties in that some villages refuse to pay leases on the lowlands, which are considered by the CFZ to be State property and for which use rights are granted to these village communities. However, this approach is seen as encouraging the participation of riparian communities in the management of these areas. On this subject, Koi Guilavogui said: "Among the *recalcitrant villages is Avillisou, which refused to pay up two years ago because it is a stronghold of the ruling RPG party. Once, he was summoned by the chief forester of Seredou, and it was the RPG section of Macenta that intervened to refuse payment of what he owed to the forestry services.*

Local elected representatives sometimes intervene to plead in favour of villages that break the law.

CHAPTER III

PARTICIPATORY MANAGEMENT AS AN ALTERNATIVE FOR RATIONAL AND SUSTAINABLE PROTECTION OF THE ECOSYSTEMS OF THE ZIAMA FOREST MASSIF

In West Africa, 1990 marked the beginning of participatory natural resource management. However, in some regions, such as East and Southern Africa, it had already begun in 1980 with claims to farmland held by a white minority to the detriment of a black majority (Maya Le Roy, 2013). From 1990 onwards, the Ziama forest began to enter the circle of participatory management. In the eyes of the forest's current managers, this marked the beginning of genuine management, which was planned following the experience of previous management under Sekou Touré.

With the advent of liberalism under the late president Lansana CONTE, awareness was raised about the need for sustainable management of Guinee's classified forests in general and those in the forest region in particular, the Ziama forest.

The initiative was taken by the late Amadou Oury BAH, National Director of Water and Forests at the time, who set out to find funding for effective management of these forests. New strategies were adopted, including the search for funding from the World Bank and the national development budget (BND), the training of managers in participatory forest management at the Tinto Water and Forestry School, and the planning and development of management plans for all the forests classified as forest regions.

Before the conservators arrived, the Ziama forest was protected by 20 supervisors. They were mainly recruited from the local villages and were given training in environmental education. These people were paid very little, but it seems that they did participate in the management of forest resources. We believe that in participatory management, if a project is to be carried out in the field, the workforce must be recruited from the local villages. This would show that these communities are not only participating in the management of the forest, but also benefiting from its advantages.

In 2013, paramilitaries were assigned to Ziama as conservators. They do have military training, but when it comes to managing forest resources, they have almost no knowledge of forestry. This risks compromising the participatory

management set up at Ziama.
The forest is the most renewable resource possible. It regenerates very easily compared with mining resources. If a well-adapted, well-coordinated and participatory strategy is drawn up by the government, and measures are taken to involve multilateral partners so that they can provide support, we will be able to monitor the method put in place.
On the other hand, if the people who live around and in these areas are interested in the effects of the resources, they must also have an interest in preserving them. When a farmer knows that by felling a palm tree he will have lost something that has belonged to him for all time and every day. If he knows that by felling a vine he will lose the next harvest, we think he will take care of it. This is the only way we will have succeeded in participative management. Barre KOIVOGUI said: "*We can't manage this forest in Conakry. It's true that we can give orders from Conakry, but when it comes to execution, those on the ground need to be well motivated and interested in the positive effects of this forest, especially the local people*.
In 1990, the PROGEFOR project tried to apply pure conservation, which consisted of cutting and uprooting the plants in the classified forest and replacing them with exported forest species. It was a way of looking only at the forest and not at people. Massa Guilavogui says, *"At first, we grew our crops peacefully by the roadside at the foot of the mountain, but one day the forestry officials came and cut it all down and threw it away, replacing it with wood that we didn't know and didn't need. We were told that we had to leave for the State"*.
This first experiment in participatory forestry in the forest was a general failure. So, for the second phase, the donor knew that it would be necessary to create a "riparian relations" division. The idea was therefore to create an agroforest, which would involve working with people who had plantations in the classified forest. This collaboration was achieved by signing a forest management contract. Under this contract, the farmer could maintain these plantations at the same time as the forest species, while the CFZ paid the costs of such maintenance.
In this collaboration, each party had the right to its species, with no one cutting for the other. These farmers were paid for the upkeep of the agroforest, just like anyone else who did the work.
For ten years, they benefited not only from the fruit of the coffee and cola trees, but also from the cost of maintaining the forest species.
Participatory management should be more than just words, it should be concrete action. It must create something to eat for the farmer, his children and his wife. The World Bank emphasises that "*participation is the process by which*

stakeholders influence and share control over the initiatives, decisions and resources that affect their development" (htt/www.worldbank.org/wbi/gouvernance/fra/about- f.html/approch).

In the community management system, control of resources is exclusively in the hands of the customary authorities (Pascal REY, 2007, ibid).

As in Tanzania, Cameroon and Ghana, where communities are fully responsible for natural resources. With the help of established regimes, they have managed to conserve their wildlife.

For the communities, the way in which agents from outside the village operate is seen as a desire to break with the existing power relationships, as Elinor OSTRON (2010, ibid, *p90*) points out: "*the regulation of common goods does not necessarily have to be imposed from the outside in a coercive manner*". The dominants will then strive to thwart the objectives of the project. For their part, the dominated will not risk going against the social consensus, which would be synonymous with exclusion and, therefore, increased vulnerability in a highly communitarian context.

According to our analyses, there are three forms of community management: the passive form, in which communities are beneficiaries of management carried out by others. In protected areas where this form of management is present, communities often have to benefit from awareness-raising sessions; co-management, a form in which communities participate in the management of natural resources through agreements. This form of participatory management gives communities real power to express their needs and assess the quality of services provided by the administration, public services and development partners. According to Malick TRAORE (2010, p 10), "*it is a space for dialogue, consultation and social accountability that enables citizens and populations to carry out participatory monitoring and evaluation of the actions intended for them and to formulate criticisms with a view to improvement*". This dynamic of monitoring and control takes into account the needs of all the players and stakeholders in the co-management system, who establish rules and agreements to maintain a degree of fairness in the distribution of tasks and the sharing of income or assets in a transparent manner. Malick Traore highlights some of the weaknesses of co-management, including the low capacity of the players, the lack of transfer of skills [41]and the community form itself. In this

[41] This management system was tried out in Mali as part of the management of three classified forests, namely Faya, Sounsan and the Monts Mandingues forest, during the first phase of the project, which lasted only four years. After evaluation, it became clear that some progress had been made, but that it was very fragile, due to the insufficient level of technical and financial ownership of the forest co-management stakeholders. We have drawn this analysis from the writings of Malick Traore (2010), entitled 'Transparence et responsabilite sociale dans la gouvernance des ressources naturelles en

form, management is genuinely carried out by the community for the well-being of the local population. Genuine community management can be severely limited by the lack of national policies to support it, regardless of the degree of involvement of donors and the potential for community projects in the region.

The current managers of the Ziama forest have adopted a participatory management approach. However, this does not obey the rules of participatory management as such, according to which the local people are fully responsible for the resources. It can be seen as a form of co-management, in which local people participate through agreements signed between them and the state. Barres KOIVOGUI had this to say: *"We are working with the local populations, we have signed contracts with them that enable them to improve their living conditions on the one hand, and on the other, we benefit from their participation in forest management"*. In Ziama, however, there is hardly any awareness-raising, and even if there is, it has no effect on the ground.

Based on our findings and the extent of community participation, we have drawn up this illustrative table.

Actors and Form	Owner of the Resources	Role of the community	Degree of local participation
Passive form	Status	Receiving benefits from forest managers, cooperating with forest managers to protect resources	Low; participation limited to essentially passive actions
Co-management (or joint management)	State, but can be decentralised or devolved	Cooperation with public authorities in the management of the protected area or resource in question	Means; depends on the rights and responsibilities conferred on local communities in a given situation
Management proper or community management	Local communities through a collective representative body	Responsible for resource management through delegated usufruct rights (rights of use) or by becoming full owners.	Eleve; communities are the main owners, decision-makers and beneficiaries.

Source: Dilys Roe, 2009. Direction: Marcel Kourouma, 2014

3-1-Participatory management: benefits for the Ziama forest

The benefits of this management include raising people's awareness of the risk of extinction of certain plant and animal species, such as rattan, which used to be abundant. Through the commitments they sign, the villages often inform the

Afrique Francophone' (Transparency and social responsibility in the governance of natural resources in French-speaking Africa), the approach of the co-management system, a space for participation or control in the management of natural resources in Mali.

head of the office of people who are not abiding by the agreement.

But another advantage of this type of management is its success. Today, when we compare satellite images from 1990-1992 with those from 2000-2001, we can see that there has been a marked change in biomass. Many of the areas that were degraded in the 1990-1992 period have been closed. This participatory management has also had advantages, such as the restoration of the massif, because it has enabled the reforestation of 26,000 hectares of forest plantations in the villages of Malaweta, Boo, Avillisou, Irie and Chefkolidou.

It should be recognised that the success of these activities is linked to the participation of these populations in the management of this resource. The communities were formed into forestry groups and worked with the conservators. They first helped to raise awareness among their relatives. Nurseries and reforestation were carried out by villagers, who earned their living by paying for contracts. This form of management was extremely important and we believe that it should be preserved.

To ensure strict compliance with the management plan, the massif has been divided into three distinct series, depending of course on the ecological importance of each series: the improvement series, the sustainable use series and the integral protection series.

The improvement series covers the areas that have been degraded since the forest was classified, and accounts for 30% of the total surface area, as well as the outskirts of the villages.

The sustainable use area is an area in which removals must be made and sold to allow sustainable management of the forest in the event of a break. It covers 27% of the total area and is located inside the forest.

The last is the integrated protection zone, located at the heart of the massif. No human activity is permitted in this area, other than research and the collection of forest seeds to supply degraded areas. It is used as a nursery for the reforestation of outlying areas. In this series, the animals, which are completely threatened by the riverside villages, have to concentrate and multiply easily so that the overpopulation can be harvested by the local people for food.

In Ziama, the areas closest to the villages are the most degraded because farmers find it necessary to farm close to the village. These areas have low vegetation. These peripheral areas are actually classified forest, but have been overexploited in the past. With the support of a number of NGOs and institutions, including PGRFOR, AFD, UNHCR, KWF and Guinea through BND, parts of these degraded areas have been reforested, with a total area of 26,000 hectares.

Today, despite the various maintenance activities carried out on these plots, the cessation of funding means that it is not possible to carry out the thinning work that is vital to the life of an artificial stand.

Diagram of the series that make up the Ziama forest and their percentage of occupation

of occupation

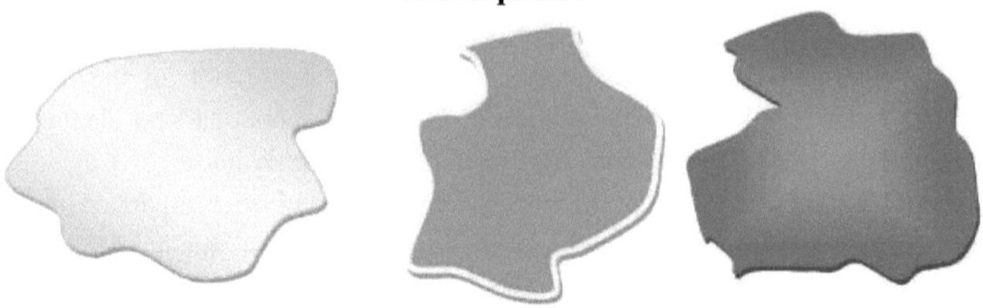

1 Improvement series 30%2 Sustainable use series 27% Integrated protection series 53

3-2-The need to integrate customary norms with the provisions of the forestry code for sustainable management of the Ziama massif

The Ziama forest is specifically under the control of the N'Nzerekore Forestry Centre. To achieve its objectives, the centre makes use of customary power. For example, the acceptance of the right of use for the local populations, the use of customs and mreurs. The harvesting of certain species in the lowlands, subsistence hunting to enable them to live as their ancestors did at the limits of the commercial activities of these resources. In return, the communities work with forestry centre staff to preserve this resource. These communities are making sure that people from outside their territory do not come to take the forest's resources for commercial purposes.

If the customs and usage rights of the riparian villages are not respected, conservation activities will fail. Koi Guilavogui had this to say: *"The traditional chiefs are also involved in the management of this forest by supporting initiatives led by the chief of the traditional healers. He had geo-located all the forests reserved for worship, as well as the watercourses where the protective genies of the villages from Seredou to Koyama are to be found"*. He produced brochures to distribute to the various villages. In this brochure, he explained the importance of preserving these places of worship.

CONCLUSION

The management plan drawn up for Ziama meets the classic objectives of conserving biodiversity and involving local people in forest management and exploitation. However, its implementation on the ground is encountering huge difficulties: lack of funding to set up programmes to raise awareness and train conservators in community management techniques, and a highly complex sociological context. Yet it is clearly defined in article 13 of the Guinean Forestry Code. However, if the forest's resources are to be preserved over the long term, concerted management is needed, i.e. an approach in which the local communities take part in the decision-making process. This implies that they should have decision-making power at all levels and access to local, national and international funding mechanisms. To achieve this, they need to be fully involved in drawing up the relevant laws and regulations, by strengthening their capacity to manage forest resources. In the light of the testimonies gathered during our field surveys, a problem arises with regard to the ownership of this forest. On the one hand, the State and, on the other, the community. Both the State and the community must make the necessary improvements to ensure the future management of the reserve's natural resources.

From a PhD perspective, we will continue with the same theme according to thc timetable below.

PERIODS	ACTIVITIES CARRIED OUT
pre Annee	Documentary research and drafting the project
2eme Year	Field work
3eme Year	Drafting and presentation of the report

BIBLIOGRAPHY

GENERAL WORKS

1-Abdelilah BAGUARD et al, 2013, " évaluation de 1 approche participative en matiere de gestion forestiere " Cas de la commune rurale Oued Ifrane, 112 p.

2-Adam SMITH, (1776), An Inquiry into the Nature and Cause of the Wealth of Nations.

3-Amadou BAH et al, (1996) Forets, politique Forestiere en Gurnee 63p

4-Alain BEITONE, (2007) Dictionnaire des Sciences economiques, 2eme edition Paris : Armand colin 252 p.

5-Banjeu GARNIER, (1958), l'agriculture Guineenne. L'information geographique, 198 p.

6-Catherine BARON, (2003), La gouvernance : débat autour d'un concept polysemique. In Droit et Societe, 400 P.

7-UNDPD, (2009), Gouvernance : 1 arbre qui cache la foret 184 p.

8-CERTU, (2001), L'analyse des systemes d'acteurs, 100 p.

9-David BOLLIER et al, (2012), les biens communs, comment cogerer ce qui est a tous? 52 p.

10-David RICARDO, (1819), Principe d'eco politique et Impot.

11-Dilys ROE et al, (2009), Gestion communautaire des ressources naturelles en Afrique : Impact, expériences et orientation future, 241 p.

12-Elinor OSTRON, (2010), Gouvernance des biens communs : pour une nouvelle approche des ressources naturelles, 300 p.

13-Fontaine CAVALERIE et al, (2004), dictionnaire de droits, 3eme edition, Paris Foucher, 511 p.

14- Garret HARDIN, (1967), the tragedy of the commons,162 p.

15- Georges PAQUET, Gouvernance : mode d'emploi, Montreal, Liber, 2008, 372 p.

15-Hermet GUY, (1998), Dictionnaire de la science Politique et des Institutions politiques; 2eme edition; Paris: Armand colin, 114 p.

16-INRETS, (2009) the concept of governance 52 p.

17-Isabelle LA CROIX et al, (2012), gouvernance : tenter une definition ; n°3, Automne, 281 p.

18-Jean pierre BOURD et Christian CHABOUD, (1993), le concept de ressources naturelles en economie, Forum Halieutique; Forum halieutique Renne 281 p.

19-Jean Pierre GAUDIN, (2001), pourquoi la gouvernance, paris: presse de sciences 11 p.

20-Jean Pierre VANDENBERG et al, (1996) anthropology and social forestry: a theoretical exploration 14 p.

21-Juhe BEAULATON, (2010), Foret Sacree sanctuaire, Paris, Karthala.
22-Jules GUILLARD, (1986), Revue du secteur forestier en Guinee.
23-Joseph STIGLITZ et al, (2007), principes d'economie moderne brochee.
24-Luc FOURNIER, (2003), Le rire des arbres. Les pleurs des forets; Outre mont, Lanctot, 250 p.
25-Martin YELKOUNI, (2004), decentralisation et dynamique des institutions dans la gestion des ressources naturelles en Afrique de l'Ouest : cas des reserves de biospheres, 53 p.
26-Marie Claire ROBIC, (1992), Du milieu a 1 environnement, pratique et representation du rapport Homme /Nature, depuis la renaissance, paris, Economica 348 p.
27-Maya LE ROY et al, (2013), la gestion durable des forets tropicales " De 1'analyse critique du concept a revaluation environnementale des dispositifs de gestion " 240 p.
28-Michel CROZIER and Erhard FRIEDBERG, (1977), 1'acteur et le systeme : les contraintes de l'action collective 500 p.
29-Ministere du Transport Public et de L'Environnement, Novembre (1997), monographie nationale sur la diversite biologique, Conakry, 148 p.
30-OIF, (2009), Feux croisises des questions de gestion des ressources naturelles par rapport au changement climatique et de renforcement des capacites des producteurs agricoles au Sahel, 240 p.
31-Olivier PETITJEAN, (2010), les biens communs : modele de gestion des ressources naturelles, 136 p.
32-Patrick TRIPLET, (2009), Manuel de gestion des aires protegees d'Afrique Francophone, 1250 p.
34- Pierre KARPE (2005), La protection des droits des collectivites autochtones sur leurs biens culturels, 250 p.
35-Pierre HAMEL and Bernard JOUVE, (2006), un modele Quebequois? Gouvernance et Participation dans la gestion publique, Montreal, 126p
36-Philippe Moreau DEFARGES, (2003) : la gouvernance, paris, presse Université de France, collection "que sais-je", 96 p.
37-Serge BAUCHETE and Pierre DE MARET, (2000), "les peuples des forets tropicales aujourd'hui" volume III, Region d'Afrique Centrale 458 p.
38-Simon OSTROWETSKY, (1997), Table ronde, colloque Geographie et Langage.
39-Thomas Robert MALTHUS, (1798), Essaie sur le principe de la population, 236 p.

MEMOIRES ET THESES 1-Charlotte Gisele KOUNA ELOUNDOU, (2012), decentralisation forestiere et gouvernance locale des forets au Cameroun : le cas des forets communales et communautaires dans la region Est, 350p, thesis in Social and Regional Geography, defended at the University of Maine.
1-Daniel LAMAH, (2009), Systemes de culture et modes d'occupation des espaces ruraux dans la zone de Gouecke, prefecture de n'Zerekore : Dynamisme et enjeux, master's thesis, 130 p.
2-Daniel LAMAH, (2013), Insertion de la cafeiculture dans les structures de production en Guinee Forestiere, thesis in geography, 490 p.
3-Daouda DIALLO, (1999). Inventaire et étude socio-economique de la roneraie naturelle de Koil, prefecture de Koundara. Memoire de DEA en sciences de l'environnement, Centre d'etude et de recherche en environnement (CERE), Universite de Conakry,120 p.
4-Florent GLATARD, (2005), Diagnostic agraire du village de Boo, Gurnee Forestiere, DEA thesis in tropical agronomy, 134 p.
5-Mamadou Cire CAMARA, (1999), Contribution a la connaissance de la Flore et de la Vegetation du Parc National du Badiar, Master's thesis in environmental science, Centre d'étude et de recherche en environnement (CERE), University of Conakry, 120 p.
6-Mohamed ELLATIFI, (2012), l'economie de la foret et les produits forestiers au Maroc : bilan et perspectives, these en science de l'environnement, 424 p.
7-Mory HABA, (2010), dynamique et évaluation des performances agronomiques et socio-economiques des plantations de palm a huiles en Guinee Forestiere, master's thesis in tropical agronomy, 34 p.
8-Pascal REY (2007), le sage et l'Etat, pouvoir, territoire et developpement en Gurnee Maritime, thesis in geography, Université de Bordeau III, 298 p.
9-Raymond DJOSSOU, (2009), Gouvernance et Performance des projets publics, 65 p.

SCIENTIFIC ARTICLES

1-Alain BERTRAND (1992). L'organisation sociale pour la destruction des forets: le cas de la Guinee-Bissau. Arbres, forets et communautes rurales, Bulletin numero *2.*
2-Djiramba DIAWARA, (2001), situation des ressources Genetiques forestieres de la Guinee, 27 p.
3-Frederique BLOT et al, (2004), "Ressources", un concept pour l'etude de relation eco-socio-systemique, Montagne Mediterraneenne, n°20, 73 p.
4-Guy DI MEO, (1987), objectivite et representativite des formations socio spatiales : de l'acteur au territoire n°537, vol 96.
5-Jean Marie KOFFI, (2013), Inegalite de droits et soutenabilite de la gestion

des ressources forestières au sud de la Cote d'Ivoire, 18 p.
6-Jean Marie MARTEL, (2002), Une demarche participative multicritere en gestion integree des forets". INFOR, vol. 40, no 3.
4-Koffi ALLINON, (2006), Societe civile togolaise face aux defis de transparence, de bonne gouvernance et d'efficacite : etat des lieux et perspectives. Question de gouvernance, les cahiers de Mande Bukari. Revue trimestrielle de l'universite Mande Bukari, No. 4, 3eme trimestre.
5-Lamine MANDIANG, (2008), "consolidating and expanding citizen communication for democratic governance in Senegal" project, 10 p.
6-Malick TRAORE, (2010), Transparence et responsabilite sociale dans la gouvernance des ressources naturelles en Afrique Francophone : l'approche du systeme de cogestion, un espace de participation ou de controle dans la gestion des ressources naturelles au Mali, 18 p.

REPORTS

FAO, (2007), Good governance of land tenure and land administration. Land Tenure Studies No 9. Rome.
1-FAO, (2009), Rapport d'Etude relatif au Dialogue sur les Forets en Afrique de l'Ouest, 250 p.
2-FAO, (2000), Approche participative, communication et gestion des ressources forestieres en Afrique sahelienne : Bilan et perspectives 120 p.
3-FAO, (2012), State of the World's Forests 66 p.
4-FAO, (1993), Projet National de Gestion des Ressources Rurales, Rapport de preparation volume 2.
5-GRIP, (2012), Ressources naturelles : conflits et construction de la paix en Afrique de l'Ouest, 36 p.
6-GRET, (2005), Construire une gestion negociee et durable des ressources naturelles renouvelables en Afrique de l'Ouest, 182 p
7-James HEATHER al, (1994), " Programme d'évaluation des trois forêts classees du sud de la Guinee " 32 p.
8-MAEF, (2007), sub-regional workshop on the sustainable management of forest resources in West Africa, 49 p.
9-Martin YELKOUNI, (2004), decentralisation et dynamique des institutions dans la gestion des ressources naturelles en Afrique de l'Ouest : cas des reserves de biospheres, 53 p.
10-Ministere de l'agriculture, de l'elevage et des forets, (2006), programme d'action National de Lutte contre la desertification (PAN/ LCD), 120 p.
11-United Nations. New-York, (2002), report of the world sleep for sustainable development, Johannesburg, South Africa, 198 p.
12-OIF, (2009), Feux croisises des questions de gestion des ressources

naturelles par rapport au changement climatique et de renforcement des capacites des producteurs agricoles au Sahel, 31 p.
13-RAC-F, (2012), Protocole de Kyoto, bilan et perspectives : vers un nouveau regime a la hauteur du defi climatique, 51 p.
14-IUCN, (2004), World Conservation Congress, 3rd Session, Bangkok, Thailand.

WEB SITES

1-http//www.acdi_cida.gc.a/INET/Image.NSF/VLUImages/Evaluation2/Sfile/ Reviewof governance_programming in CIDA_EN.pdf (page consulted on 04 April 2014.
2-http//www.worldbank.org/wbi/gouvernance/fra/about-f.html/approch, page consulted on 04 April 2014.
3-draaf.pays-de-la-loire.agriculture.gouv.fr/.../Tome_1_-_chapter_11_cle
4-htt//eegeosociale.free.frirenne2006 ;"De l'usage de la notion d'acteurs en Geographie" (How social geography, after winning the war of sciences, can build peace), page consulted on 09 April 2014.
5-htt://www.ecosociosysteme.fr/ressources naturelles-html (page consulted on 08 April 2014).
6-htt://www.crpf.fr/bretagne/doc-gestion-durable/fonction_foret.html (page consulted on 10 April 2014).
7-http://www. guinee. gov. gn/1 bienvenue/discours.htm (site consulted on 01/10/ 2014.
8-htt://www.ofme.org/crpdf (page consulted on 10 April 2014).
9-http://eur. lex.europa.eu/LexUriserv/site/en/com/2001/com2001 0428enO1.pd f (page consulted on 04 April 2014).
10-WWW. Icra - edu .org

APPENDIX

Annex 1: Summary table of classified forests in the Gurnee Forestiere

Forest names	Area in hectares (ha)	Degrees of degradation (in %)	Altitude (m)
Dere	8 920	90	-
Bero	26 850	75	1 200
Diecke	59 143	30	595
Pic de Fon	25 600	35	1 656
Banan	12 000	40	-
Yonon		-	0
Ziama	119 019	27	1 387

Source: CFZ, realise par le candidat 2014.

Annex 2: Complaint registered by the Ziama Branch Manager about damage caused by elephants from 1999-2014

N0	Villages	Name of victims	Degats	Date of damage
1	ZOBOROMA	MORIBA GUILAVOGUI	RICE AND COFFEE	
		AWARA GROVOGUI	RICE AND COFFEE	
		SOUARE TOUPOU	BANANES	
		OUA PIVI	RICE	
		GUILE GUILAVOGUI	MANIOC	20/11/199
2 3	ERIE BALASSOU	ZEZE BILIVOGUI	RICE	2003
		GBABE BEAVOGUI	RICE	2010
		MORY SOVOGUI	RICE	
		MOUSSA TRAORE	RICE	
		KABINET TRAORE	RICE	
		CEZE GUILAVOGUI	COOKING and RICE	
		SEKOU BEAVOGUI	RICE	
		GABRIEL KOIVOGUI	RICE	
		FATA CAMARA	RICE	
4	AVILISSOU	KOIKOI BILIVOGUI	RICE	
		MOUSSA TRAORE	GRENIER	
		FALIKOU SAGNO	RICE, TARO, GRANARY	
		MALICK SANGARE	TARO and BANANES	2013
		ABOU DIALLO	BANA NES, TARO and CAFE	
		LAMINE KEITA	BANANES	
		MORIFING TRAORE	BANANES	
		SIBA ONIVOGUI	BANANES and COFFEE	
		MAMADI KOUROUMA	CACAO and TARO	
		BANGALY TRAORE	RICE AND COFFEE	2014

Source: CFZ, Ziama branch, realised by the candidate, August 2014

Annex 3: SUMMARY OF CLASSIFIED FORESTS IN GUINEA

Regions	Area (ha)	Number of classified forests	Classified area (ha)	Largest forest in the region	Percentage of coverage
Low (thin	3,620,800 ha	32	112 068 ha	Botokoly 23 000 ha	3,09
Average GuinCe	6 360 800 ha	64	413 638 ha	N'dama 67,000 ha	6,50
Haute GuinCe	9 666 700 ha	26	333 723 ha	Kouya 67 400 ha	3,45
GuinCe ForestiCre	4,937,400 ha	8	322 704 ha	Ziama Massif 119019 ha	6,53
TOTAL	**24,585,700 ha**	**130**	**1,182,133 ha**	**269,700 ha (22.81% of the classified area)**	**4,80%**

Appendix 4: MAINTENANCE GUIDE

Interview guide

Interview guide for managers of government departments.

Name and surname of the investigator :

Identification number :

Survey date: from /.... / to /.... / 2014

Start of 1 interview :

End of 1 interview :

Duration of interview :

Identification of the interviewee

Full name :

Gender :

Age :

Marital status :

Service :

1-What is the role of public authorities in the management of protected areas?

2-What role does customary power play in the management of protected areas?

3-What measures does the State plan to take to prevent the destruction of classified forests?

4-What is your analysis of the villages at the heart of the massif?

5-What role does your institution play in the management of this forest?

6-What is the relationship between the forestry centre and the local population?

7-Are the laws governing the regulation of the forest respected by the companies? populations? yes or no

If not, why not

8 Are there any conflicts between your institution and the local population? yes or no

9 If a conflict arose, how would you deal with it?

10-Do you run campaigns to raise awareness of the challenges of deforestation in your country?

these villages? yes or no

If so, how

If not, why not?

11-Do local people benefit from compensation for damage caused by the animals in the fields? yes or no

If so, how much?

12-Are the communities involved in drawing up the laws that regulate the forest? yes or not

If yes,

what?

If not,

Why

13-Do you think that 1 Etat is better p1acë for дёгег this forest? yes ______||or No |

14-Is management carried out exclusively by the State or is it shared?

15-What management systems do you envisage for the sustainable management of this reserve?

16-Are NGOs involved in the management of this massif? yes or no

If so, how many were involved between 1942 and the present day?

17-What challenges are you facing?

18- Do these communities cause damage to the forest? yes or no

if so, how

19-Who owns this forest today? And why?

a- *Community*

b- *State*

c- *Private*

INTERVIEW GUIDE FOR THE GENERAL PUBLIC

Full name of the investigator:

Gëographic CoordoniK'e (GPS)

Identification number :

Survey date: from /.... / to /.... / 2014

Dëbut de 1 interview :

End of 1 interview :

Duration of interview :

Identification of the interviewee

Full name :

Gender :

Age :

Marital status :

Profession :

Village;Neighbourhood: ; CRD :

1- How was the forest managed before it was classified?

2- Who were the first inhabitants of your village, and where did they come from?

How is land appropriated in your community?

1. *Droit de Hache* |
2. *by heritage*
3. *By donation*
4. *sale*

3-was management governed by laws? yes I I or no I I

If so, which ones?

4- If, for example, a resident were to break this rule, what would be the consequences?
a- *struck by the wrath of genius* ____||
b- *pays a fine to the community*
5-How do you relate to the forest?
6- What are the different activities you used to do?
a- agriculture

b- breeding	
c- *fishing*	
d-commerce	

7-Are boundaries drawn to prevent penetration of the forest? yes __|| or no IZZI
8- Does physical violence occur between you and forest rangers? yes _____II or no ______||
If so, why?
Because :
a-you kill animals ______II
b- you cut the wood ______||
c-or you break the rules _____||
9- Does community violence sometimes occur between you? yes □ or
If so, what are they?
10-What are the ways of resolving these *a-customary* conflicts ?
b- *judicial*
d- *family*
11- Do you know the content of the documents governing the forest? yes I lou no I I
12- Do you experience any problems during the growing season? yes or no IZZI
a- animals ravage the fields |^ZI
b- forest rangers prevent the extension of fields ____||
13- Are there any NGOs that support you in activities outside the scope of your work?
l'agriculture? yes ____|| or no ||
If so, which ones?
14- Who owns this forest today? And why?
a- *community* b- *state* c- *private*

MIX
Papier aus verantwortungsvollen Quellen
Paper from responsible sources
FSC® C105338

Printed by Books on Demand GmbH, Norderstedt / Germany